Springer Tracts in Modern Physics 76

Ergebnisse der exakten Naturwissenschaften

W0235255

Editor: G. Höhler
Associate Editor: E. A. Niekisch

Editorial Board: S. Flügge J. Hamilton F. Hund
H. Lehmann G. Leibfried W. Paul

Manuscripts for publication should be addressed to:

G. Höhler
Institut für Theoretische Kernphysik der Universität Karlsruhe
75 Karlsruhe 1, Postfach 6380

Proofs and all correspondence concerning papers in the process of publication should be addressed to:

E. A. Niekisch
Institut für Grenzflächenforschung und Vakuumphysik
der Kernforschungsanlage Jülich, 517 Jülich, Postfach 365

Hans Ullmaier

Irreversible Properties of Type II Superconductors

With 67 Figures

Springer-Verlag Berlin Heidelberg GmbH 1975

Dr. Hans Ullmaier

Institut für Festkörperforschung der Kernforschungsanlage Jülich
517 Jülich, Postfach 1913

ISBN 978-3-662-15859-3   ISBN 978-3-540-37963-8 (eBook)
DOI 10.1007/978-3-540-37963-8

Library of Congress Cataloging in Publication Data. Ullmaier, Hans, 1936. Irreversible properties of type II superconductors.
(Springer tracts in modern physics, vol. 76). Includes bibliographical references and index.
1. Superconductors, Type II. 2. Irreversible processes. I. Title. II. Series. QC1. S 797. vol. 76.
[QC 612. S8]. 539'. 08s. [537.6'23]. 75-26967.

© by Springer-Verlag Berlin Heidelberg 1975.
Originally published by Springer-Verlag Berlin Heidelberg New York in 1975.
Softcover reprint of the hardcover 1st edition 1975

Dem Andenken an meinen Vater gewidmet

Preface

Although Kamerlingh Onnes discovered superconductivity in
1911, it was almost exactly half a century later before technological applications of this phenomenon began to emerge. The
reason for this long span of time was the tendency to learn more
about the nature of superconductivity by investigating mainly
pure metals (now called type I superconductors), in which superconductivity is destroyed by very low magnetic fields. In 1961
however, Kunzler and his coworkers found that the compound Nb_3Sn
was able to carry high current densities in magnetic fields of
100 kG and more without losses. This and the discovery of other
high field materials ("hard" type II superconductors) opened the
possibility of creating and maintaining strong magnetic fields
with negligible power input.

Superconducting coils are being and will be used as laboratory magnets, in bubble chambers and particle accelerators, in
large turbogenerators, for high speed trains, etc. Hopefully the
most important future application will be the confinement of the
plasma in large thermonuclear devices which eventually shall lead
to fusion reactors.

Many scientists have contributed to the development of hard
superconductors - sometimes making mistakes and correcting them -
adding bricks, some marked with their names and many others ano-
nymous. Necessarily this monograph cannot cover the history of
their efforts. Instead I have tried to present ideas and models
that provide the best means known today for understanding the
physics of hard superconductors and connecting it with the fields
of application. In this way the book may be of help to physicists,
metallurgists, and engineers working in these fields.

This monograph was written during a stay at the Oak Ridge National Laboratory and I am very indebted to the members of the Radiation Effects Group in the Metals and Ceramics Division for their kind and patient hospitality. I also wish to thank the Kernforschungsanlage Jülich for the exchange arrangement which made my stay in Oak Ridge possible. Valuable discussions with many colleagues in Jülich and elsewhere, especially with Dr. G. Antesberger, Prof. R. Labusch, and Dr. J. Schelten are gratefully acknowledged. Dr. C.C. Koch, Dr. G. Lippmann, and Dr. R.M. Scanlan informed me about some of their results prior to publication.

My particular gratitude is due to my staunch friend R.H. Kernohan for numerous helpful suggestions, for his careful reading of the manuscript, and for converting many awkward expressions into readable English. Last but not least I am most grateful to Professor W. Schilling for his friendship and for his continuous encouragement and support.

Oak Ridge, and Jülich
January 1975   Hans Ullmaier

Table of Contents

1. Introduction

In type II superconductors which are free of extended
lattice defects, that is, "ideal" type II superconductors, the
flux line assembly present in the mixed state will begin to
move whenever a force acts on it. The velocity of this flux
flow is determined by a balance between the applied force and
a viscous drag force. The motion of vortex lines generates
electric fields in the superconductor resulting in dissipative
and galvanomagnetic effects. Since a transport current perpendicular to the field direction is equivalent to a force on the
flux line system, ideal type II superconductors in the mixed
state are not able to carry transport currents without losses.

The situation is different in the presence of extended
lattice defects (dislocations, grain boundaries, precipitates,
etc.) which act as pinning centers for the flux lines. The pinning centers prevent the flux lines from moving until the
applied force density exceeds a critical value. The existence
of such a critical pinning force density gives rise to macroscopic flux density gradients and irreversible magnetic behavior,
i.e., the number of vortices present differs from that in thermodynamic equilibrium and strongly depends on the magnetic history of the sample. Such irreversibly behaved type II superconductors are often called hard superconductors[+] in analogy
to mechanically or magnetically hard materials. All technologi-
cally useful materials require high current carrying capacity
in high magnetic fields, and therefore must be hard superconductors. Actually most type II materials are more or less hard

---

[+] Sometimes the somewhat misleading terms type III or high field superconductors are used in the literature.

superconductors since each metal or alloy usually contains
enough imperfections to cause pinning and only in the last few
years could sufficiently perfect samples be prepared to show
almost reversible ("ideal") behavior.

The irreversibility of most alloy samples investigated
between around 1930 and 1960 was probably the main reason that
type II superconductivity was so late in being recognized. The
only experiments on fairly reversible samples by SHUBNIKOV et al.
(1937) were obscured by extensive data on non-ideal two-phase
materials which were interpreted by the "sponge" model (MENDELS-
SOHN 1935). Also, progress was inhibited by World War II and by
the fact that Shubnikov's work was published in Russian. Although Abrikosov's theory already identified alloy superconductivity with the case $\kappa > 1/\sqrt{2}$ of the Ginzburg-Landau theory in
1952, it took another ten years until GOODMAN (1961) suggested
that the very high critical fields of Nb_3Sn discovered by KUNZ-
LER et al. (1961) might be associated with type II behavior.
Shortly thereafter several experimental investigations provided
evidence in support of this view and clearly showed that hard
superconductors are type II materials where the flux line system is pinned by various lattice imperfections.[+] The concept
of the "critical state", introduced by H. LONDON (1962), BEAN
(1962), and KIM et al. (1962), provided a phenomenological description of the macroscopic behavior of hard superconductors
and reduced the variables to a single material-sensitive pro-
perty, the pinning force P_v per unit volume. The first microscopic explanations of this behavior are those of GORTER (1962)
and ANDERSON (1962) who suggested a spatial variation of the
free energy as a cause for pinning.

In the years following 1962 the interaction forces, K, of
a single vortex with several specific types of defects were calculated. However, a simple summation of these forces over the
pinning centers contained in unit volume usually gives much
higher macroscopic forces P_v than found experimentally. This
discrepancy was explained by YAMAFUJI and IRIE (1967) and LA-

---

[+] An almost complete presentation of this early work on type II superconductivity is found in the Proceedings of the Intern. Conf. on the Science
of Superconductivity held at Colgate University, Hamilton, N.Y., Aug. 26-29,
1963, published in Rev. Mod. Phys. 36, (1964).

2

BUSCH (1969a) who first pointed out that the summation of individual vortex-defect interactions must involve the elastic interaction of the flux line lattice as a whole with the statistical array of pinning centers.

Besides these activities in improving the understanding of the physics of pinning, considerable progress was made in understanding and overcoming the problem of thermomagnetic instabilities in hard superconductors. These instabilities which are caused by dissipative movements of vortices past their pinning barriers are of great technological importance since they can be the limiting factor for the performance of superconducting coils. Modern commercial multifilament wires are "intrinsically stable" and allow the design of powerful magnets with predictable performance.

Whereas the properties of flux lines in ideal type II superconductors will be covered in a forthcoming review article this monograph deals with those aspects which are characteristic of hard superconductors. In the following the subjects of forces on flux lines (Chapter 2) and the connection between macroscopic force densities and individual interaction forces between single flux lines and defects (Chapter 4) are treated in some detail. Chapter 3 gives a short survey of possible pinning mechanisms for which theoretical expressions for the maximum interaction force are derived. They can be compared with experimental data on model systems (Chapter 5) which permit the determination of the force due to a specific type of defect. Chapter 5 also contains short sample data of some commercial alloys used in superconducting magnet coils. A brief review of dissipative effects which may lead to thermomagnetic instabilities is given in Chapter 6, followed by a description of experimental techniques applicable to pinning force measurements (Chapter 7). The description of phenomena in hard superconductors is of course based on properties of flux lines in reversible materials which are treated in the reviews by DE GENNES (1966), ST. JAMES et al. (1969) or FETTER and HOHENBERG (1969). Since sometimes they may not be on hand a short outline of reversible properties is given in appendix A (Chapter 8). Finally appendix B contains a list of symbolds and their dimensions in MKSA-units which are used throughout the book.

Since its discovery some fourteen years ago, the phenomenon of vortex pinning in hard superconductors has been and still is the subject of intense experimental and much less intense theoretical research. In this book I have tried to summarize the results of these activities and to describe our present ideas of the physics of hard superconductors. Due to the limited size of this book most subjects are treated rather concisely. I did not endeavor to duplicate the excellent comprehensive monograph by CAMPBELL and EVETTS[+], but I strongly recommend it for delving more deeply into the subject and for a full bibliography.

[+]A.M. CAMPBELL and J.E. EVETTS, Critical Currents in Superconductors, Taylor and Francis Monographs on Physics (ed. B.R. Coles and Sir Neville Mott), London 1972 (reprinted from Advances in Physics 21, 199, March 1972).

2. Forces on Flux Lines

A flux line consists of super-electrons moving with a certain density and velocity distribution around the center of the line. This means that forces on a flux line will actually be experienced by its super-electrons and can therefore be caused by magnetic or electric fields, gradients in the electron chemical potential, or accelerations. In order to calculate the different individual interactions, a detailed knowledge of the microstructure of the flux lines is necessary.

However, in dealing with macroscopic quantities (like transport currents, flux density gradients, etc.) it is sufficient to consider averaged values and employ an electrodynamic and thermodynamic description of the phenomena.

For the following it will therefore be practical to divide the forces into two groups, depending on their interaction range compared to the characteristic lengths of the flux line arrangement. The first group consists of forces with ranges smaller or about equal to the penetration depth λ. These will be designated as local forces, K, since they act directly only on one or a few individual vortices. A typical example of such forces is the attraction between a small normal conducting precipitate and a flux line core.

The second group consists of forces which act on a large number of flux lines simultaneously, as e.g., forces caused by a transport current flowing through the bulk of a sample. Such "macroscopic" forces are usually expressed in terms of force densities P.

2.1 Macroscopic Forces

In order to define macroscopic quantities which are suitable for the description of the mixed state we consider a system of straight vortices in an ideal type II superconductor in an external longitudinal field H_o.[+]

In Fig. 1 the changeover from the microscopic to the macroscopic description is shown schematically. The microscopic

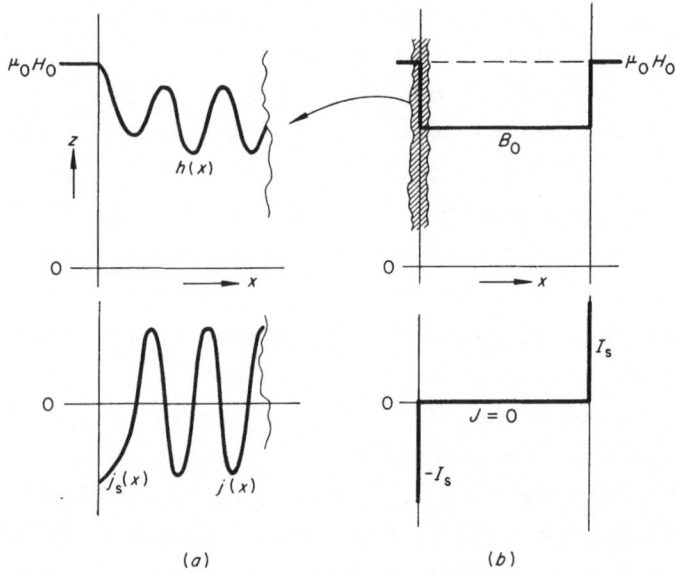

(a)   (b)

Fig.1   (a) Microscopic and (b) macroscopic description of the mixed state in a long slab in a longitudinal magnetic field H_o. The microscopic field distribution $h(x)$ is due to the superposition of the fields of vortices with equilibrium density n_o. A surface current with density j_s maintains the difference between $\mu_o H_o$ and $\langle h(x) \rangle$. In the macroscopic picture these quantities are replaced by the average flux density B and the surface current per unit length I_s. The average current density J in the bulk is zero.

[+] If not stated otherwise, this simple "one-dimensional" geometry will always be used. It has the advantage that all equations will take a very simple form and that the physics of the phenomena is not confounded by geometrical effects (caused by demagnetization factors, flux line curvatures, etc.). In the one-dimensional case the directions of magnetic fields, currents, and driving forces are always perpendicular to each other and we may therefore disregard the vector properties of these quantities. The general vector expression for the driving force is given by Eq. (2.26) and further discussed in Chapter 7.

field distribution h(x) is replaced by the flux density B, and the local current density j(x) by the average current density J (which is zero in this special case), and a surface current I_s per unit length takes the place of the surface current density $j_s(x)$.

$$B = \langle h(x) \rangle = n\phi_0 \qquad\qquad (2.1)$$

$$J = \langle j(x) \rangle = \frac{1}{\mu_0} \frac{dB}{dx} \qquad\qquad (2.2)$$

$$I_s = \int j_s(x)\,dx = H_0 - B/\mu_0 \qquad\qquad (2.3)$$

The isothermal equilibrium mixed state structure for a given external field H_0 is reached if the Gibbs free energy density

$$g = f - B \cdot H \qquad\qquad (2.4)$$

is a minimum, i.e., if

$$\frac{dg}{dB} = \frac{df}{dB}\bigg|_{B = B_0} - H_0 = 0 \ . \qquad\qquad (2.5)$$

The equilibrium flux density B_0 can be calculated from Eq. (2.5) if the free-energy density f as a function of B is known. In many cases and especially for hard superconductors it is useful to define a general thermodynamic field H (which then also exists within the superconductor, see dotted line in Fig. 1b) by

$$H \equiv \frac{df}{dB} \qquad\qquad (2.6)$$

H defined in this way has the correct electrodynamic properties usually associated with H, e.g., in vacuum $f = B^2/2\,\mu_0$ and $H = df/dB = B/\mu_0$.

Having now introduced the quantitites for the macroscopic description of the mixed state we can consider the forces for the case in Fig. 1. In the surface region of the slab the pressure due to the external field, F_0, is opposed by a force due to the surface currents F_S and the internal pressure of the vortex system F_M:

$$F_o = \frac{\mu_o H_o^2}{2} = F_s + F_M = \frac{1}{2}\left(\mu_o H_o^2 - \frac{B_o^2}{\mu_o}\right) + \frac{B_o^2}{2\mu_o} \qquad (2.7)$$

Since B_o (and therefore F_M) is independent of position, dF_M/dx is zero and no macroscopic net force is transmitted to the crystal lattice. This rather trivial result was of course expected since the situation just corresponds to the equilibrium state.

2.1.1 Flux density gradients

In the above case of a uniform flux density the repulsive interactions between a flux line and all its neighbors cancel each other and no net force occurs (Fig. 2a). This situation is changed in the presence of a flux density gradient where we expect a driving force on the vortex arrangement in a direction opposite to that of the flux density gradient (Fig. 2b).

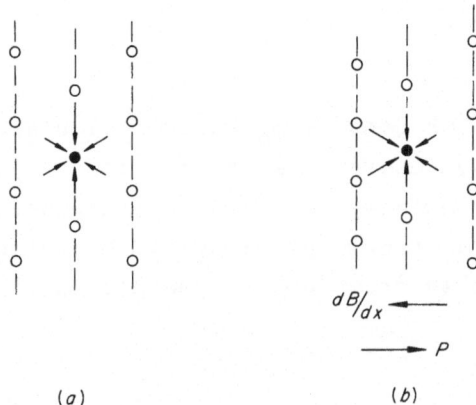

(a)   (b)

Fig.2 Local forces experienced by an individual flux line (full circle) from its neighbors (open circles) for (a) a uniform flux density and (b) a flux density gradient dB/dx. In this schematic diagram the difference in the flux line distances is greatly exaggerated. In reality this difference is either much smaller or the density gradient ist produced by dislocations (see Section 2.2.2).

In order to derive a relation between the driving force per unit volume, P_D, and the flux density gradient we consider a similar slab geometry as in Fig. 1, but with the ideal material

8

now replaced by a hard superconductor. We will assume that the
hard superconductor is characterized by a certain maximum pinning force density P_V (without, at the present time, considering
the sources for flux pinning). We will further assume that P_V
is independent of B. Although this is usually not the case it
simplifies the discussion without loss of generality for the
resulting formulas.

Fig. 3 shows the macroscopic field and current distributions in a hard superconductor. These distributions result from
the combined influence of a longitudinal external field H_M and
a transport current I in y-direction per unit length of the
slab in z-direction. The situation in Fig. 3 corresponds to the
so-called critical state where the driving force density P_D is
just balanced by the pinning force density P_V.

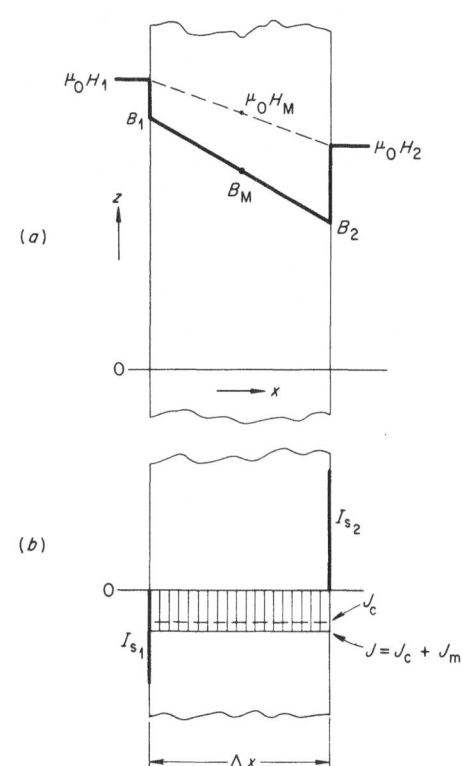

(a)

(b)

Fig.3 Hard superconducting slab under the combined influence of an external field (in z-direction) and a
transport current (in y-direction):
(a) macroscopic field distribution
and (b) current distribution.

Making use of the condition that the sum of all forces acting on the slab must be zero, we obtain

$$F_e = F_{S_1} - F_{S_2} - P \, \Delta x \tag{2.8}$$

where F_e is the net external force per unit area of the slab which is balanced by surface forces F_S and a bulk force $P \cdot \Delta x$. The indices 1 and 2 refer to the left and the right surfaces of the slab, respectively. Using Eq. (2.7) we may rewrite Eq. (2.8)

$$\frac{\mu_o}{2} (H_1^2 - H_2^2) = (\frac{\mu_o H_1^2}{2} - \frac{B_1^2}{2\mu_o}) - (\frac{\mu_o H_2^2}{2} - \frac{B_2^2}{2\mu_o}) - P \, \Delta x \tag{2.9}$$

or

$$P = - \frac{1}{2\mu_o} \frac{B_1^2 - B_2^2}{\Delta x} \tag{2.10}$$

Equation (2.10) shows that the bulk force density is opposite and equal to the gradient of the "magnetic pressure". The force density P is transmitted to the bulk by two contributions

$$P = P_D + P_L \tag{2.11}$$

P_D is the driving force density which acts on the flux line system whereas the lattice force density P_L is transmitted directly to the crystal lattice.

At first sight the occurrence of the second term in Eq. (2.11) is somewhat surprising since, by analogy with the normal state, one might expect the driving force density to be equal to the Lorentz force

$$- \frac{B_M}{\mu_o} \frac{\Delta B}{\Delta x} = - \frac{B_1 + B_2}{2\mu_o} \frac{B_1 - B_2}{\Delta x} = P_D \tag{2.10a}$$

which would mean that $P_L = 0$. However, a simple Gedankenexperiment introduced by CAMPBELL and EVETTS (1972, p.232) illustrates that $P_L \neq 0$ for $\Delta B / \Delta x \neq 0$. They considered a slab of an ideal type II superconductor in which the Ginzburg-Landau parameter κ

10

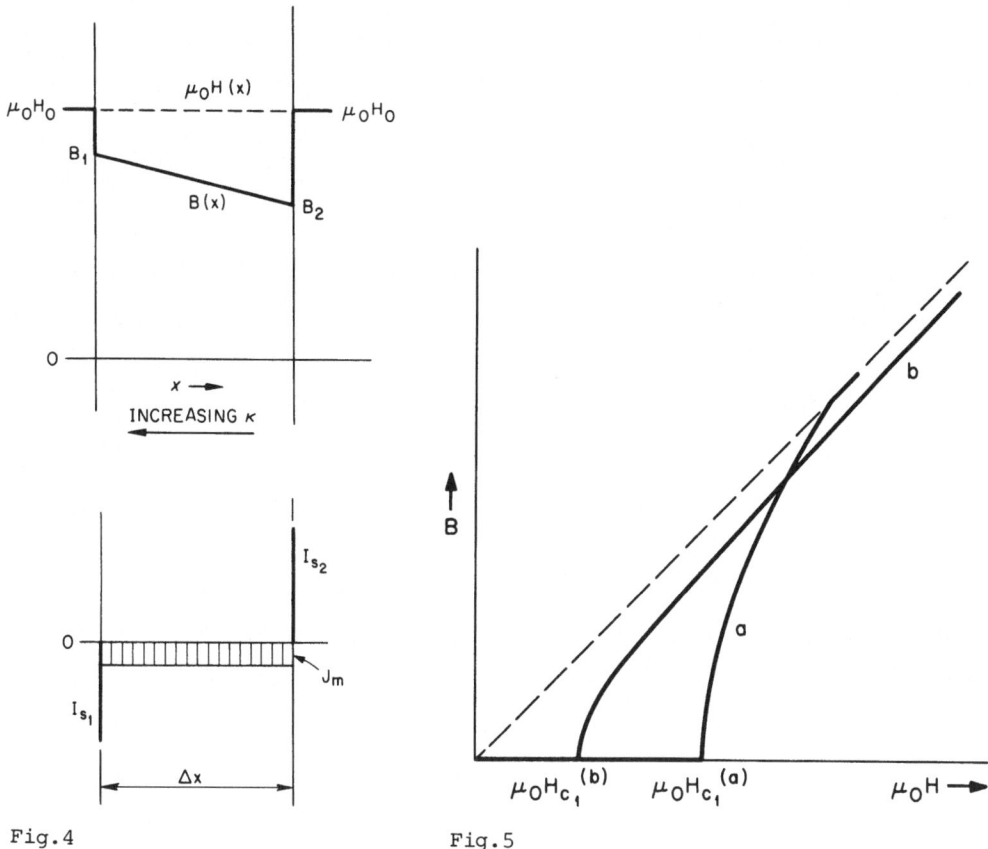

Fig.4   Fig.5

Fig.4   Ideal type II superconductor with Ginzburg-Landau parameter κ varying smoothly across the sample thickness. The bulk current density J_m is
proportional to the gradient in the magnetization M = B - μ_oH.

Fig.5   Macroscopic flux density B as a function of magnetic field H for
ideal type II superconductors with a (a) low and (b) high Ginzburg-Landau
parameter. For the high-κ material $dB/\mu_o dH \approx$ 1 (except very close to H_{c1}).

varies smoothly across the sample, i.e., κ = κ(x) (Fig. 4). Although the external field H_o is the same, the flux densities
near the left and right surface are different because of the
different equilibrium relations B = B(H_o) (see Fig. 5). Since
there is no external current through the slab we have

$$0 = I_{S_1} - I_{S_2} + J_m \Delta x .$$ (2.12)

The density J_m of the magnetization current is then obtained by
using Eq. (2.2) as:

$$J_m \Delta x = (\frac{B_1}{\mu_o} - H_o) - (\frac{B_2}{\mu_o} - H_o) = \frac{1}{\mu_o}(M_1 - M_2) \tag{2.13}$$

where M_1 and M_2 are the magnetizations at the left and right side, respectively. J_m corresponds to a volume force $P_L = -B_M J_m$ which is supported by the crystal lattice. $P_L \Delta x$ just balances the difference of the surface forces fulfilling the condition that the net external force on the slab in Fig. 4 is zero, i.e.,

$$0 = F_{s1} - F_{s2} - P_L \Delta x =$$

$$(\frac{\mu_o H_o^2}{2} - \frac{B_1^2}{2\mu_o}) - (\frac{\mu_o H_o^2}{2} - \frac{B_2^2}{2\mu_o}) - \frac{B_1 + B_2}{2}\left[(\frac{B_1}{\mu_o} - H_o) - (\frac{B_2}{\mu_o} - H_o)\right] \tag{2.14}$$

In the general case where the external field is not the same on both sides of the slab (Fig. 3a), Eq. (2.13) is modified to

$$J_m \Delta x = \frac{1}{\mu_o}(M_1 - M_2) = (\frac{B_1}{\mu_o} - H_1) - (\frac{B_2}{\mu_o} - H_2) \tag{2.15}$$

which yields a lattice force

$$P_L \Delta x = -\frac{B_1 + B_2}{2}(\frac{B_1}{\mu_o} - H_1) - (\frac{B_2}{\mu_o} - H_2) \tag{2.16}$$

Combining Eq. (2.16) with (2.10) and (2.11) we finally obtain

$$P_D = P - P_L = -\frac{1}{2\mu_o}\frac{B_1^2 - B_2^2}{\Delta x} - \frac{B_1 + B_2}{2\Delta x}\left[(\frac{B_1}{\mu_o} - H_1) - (\frac{B_2}{\mu_o} - H_2)\right]$$

$$= -B_M \frac{H_1 - H_2}{\Delta x} \tag{2.17}$$

Going to the lim $\Delta x \rightarrow 0$, the bulk force balance of Eq. (2.11) is given by

12

$$P = -B \, \frac{1}{\mu_o} \frac{dB}{dx} = P_D + P_L = -B \, \frac{dH}{dx} - B \, \frac{1}{\mu_o} \frac{dM}{dx} \tag{2.18}$$

Since $dB_o/\mu_o dx$, dH/dx, and $dM/\mu_o dx$ are equivalent to current densities, Eq. (2.18) can be written in the more familiar Lorentz type form

$$B \cdot J = B \cdot J_c + B \cdot J_m \tag{2.19}$$

This leads to the concept of a division of the total bulk current density $J = \langle j(x) \rangle$ (see Eq. 2.2) into a transport current density J_c and a magnetization current density J_m. Although physically this division is rather irrelevant since both J_c and J_m are supplied by the same charge carriers, it is helpful in understanding the different force transferring mechanisms in the superconductor.

In the case of Fig. 3 the vortex arrangement remains stationary as long as the driving force P_D is smaller or just equal to the maximum pinning force density, i.e., the critical state is reached if

$$P_V = -P_D = B \, \frac{dH}{dx} = B \, \frac{dH}{dB} \frac{dB}{dx} = BJ_c \tag{2.20}$$

As soon as $|P_D| > |P_V|$, flux flow occurs and an additional force will appear in Eq. (2.11) which is due to the viscous resistance of the moving vortices (see Chapter 6). J_c is usually called the critical current density since it determines the maximum transport current that a hard superconductor can carry without losses.

In order to examine to what extent the two terms in Eq. (2.11) contribute to the total bulk force density we consider the ratio

$$\frac{J_m}{J_c} = \frac{\mu_o^{-1} \, dB/dx - dH/dx}{dH/dx} = \frac{1}{\mu_o} \frac{dB}{dH} - 1 \tag{2.21}$$

This ratio is independent of the pinning force density and is determined by equilibrium properties of the superconductor only, i.e., for a given material the following statements concerning P_L and P_D are valid for all pinning strengths. Equation (2.21) shows that the lattice force term P_L will be important only if the slope $dB/\mu_o dH$ differs significantly from unity which is the case for fields close to H_{c1} and/or low-κ materials (see Fig. 5). In all other cases P_L is very small compared to P_D and can be neglected for most practical purposes.

Historically the important relation (2.20) was derived first by FRIEDEL et al. (1963) employing a thermodynamical approach. They considered an area A with N straight vortex lines as a system of particles. The analogue of pressure (per unit vortex length) is

$$p = - \left.\frac{\partial (Af)}{\partial A}\right|_T = - f - A \frac{\partial f}{\partial A} \qquad (2.22)$$

If the number $N = A \cdot B/\phi_o$ of flux lines is constant, we have $BdA + AdB = 0$ and therefore

$$p = - f + B \frac{\partial f}{\partial B} \qquad (2.23)$$

Hence the driving force per unit volume $P_D = - dp/dx$ is given by

$$P_D = - (- \frac{df}{dB} + \frac{df}{dB} + B \frac{d^2 f}{dB^2}) \frac{dB}{dx} \qquad (2.24)$$

According to Eq. (2.6) $df/dB = H$, i.e.,

$$P_D = - B \frac{dH}{dB} \frac{dB}{dx} = - B \frac{dH}{dx} = - BJ_c \qquad (2.25)$$

which is identical to Eq. (2.20) and again shows that only the portion $(dH/dB) \cdot (dB/dx)$ of the total flux density gradient contributes to the driving force P_D.

Since the thermodynamic quantities used in deriving Eq. (2.20) and (2.25), respectively, are scalar functions it is difficult to generalize the result for the three-dimensional case. Different approaches have been used by KOPPE (1965a), JOSEPHSON (1966) and EVETTS and CAMPBELL (1966) to overcome this difficulty. We will not go into the details of these calculations but will only give the final result for the driving force density

$$- \underline{P}_D = \underline{B} \times \text{curl } \underline{H} = \underline{P}_V \qquad (2.26)$$

which again takes the form of a Lorentz force if curl $\underline{H}$ is identified with the transport current density J_c rather than the total bulk current density J.

Some applications of Eq. (2.26) will be discussed in Section 7.1. Here we will only mention the most simple case of a long cylinder in a longitudinal field since this geometry is preferentially employed in experimental investigations of type II superconductors. Expressing Eq. (2.26) in cylindrical coordinates we obtain

$$P_D = - B \frac{dH}{dr} = - B \frac{dH}{dB} \frac{dB}{dr} \qquad (2.26a)$$

which is as simple as Eq. (2.20) derived for the one-dimensional slab geometry. Equation (2.20a) is the basis for the determination of pinning force densities by measuring the critical flux density gradient dB/dr (see Section 7).

Equation (2.20) or (2.26), respectively, enable us to calculate the maximum flux density gradient (or critical current density) of a certain material if the pinning force density P_V is known. This information is, however, not sufficient to predict the entire flux distribution within a given specimen since the flux density distribution in a hard superconductor does not represent the equilibrium state and is therefore strongly dependent on the way in which it was created. In order to illustrate this influence of the magnetic history we discuss two simple examples, both being of interest for the interpretation of experimental data.

The first example deals with flux distributions in a long cylinder which is subject first to a slowly increasing and then

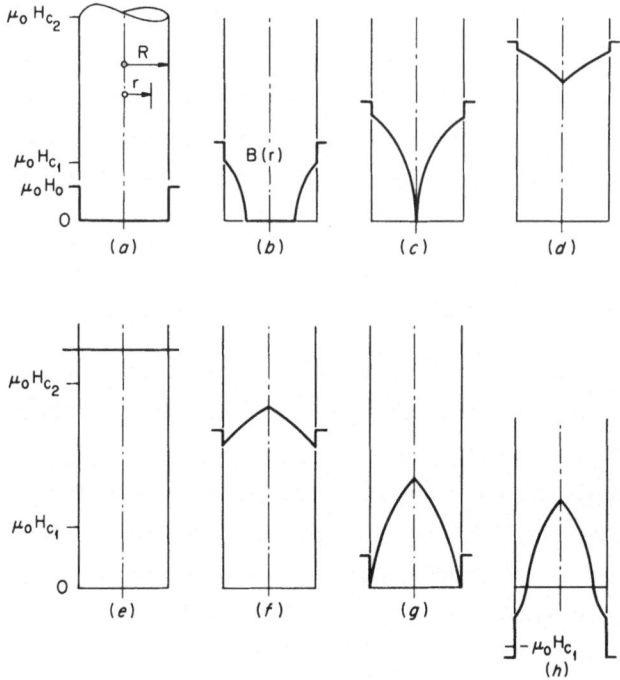

Fig.6 Macroscopic flux density distribution B(r) in a hard superconducting cylinder in increasing [(a) to (e)] and decreasing [(e) to (h)] external fields (schematic).

slowly decreasing longitudinal external field, a cycle which is usually carried out during the recording of a magnetization curve (see Section 7.2). For increasing external fields $H_o < H_{c1}$ no flux lines are present in the bulk of the sample (Fig. 6a). As soon as H_o exceeds H_{c1} the Meissner state becomes unstable and flux lines are created near the surface of the specimen. In an ideal material they would distribute homogeneously across the whole sample cross section according to the equilibrium condition $B_o = B_o(H_o) = $ constant (Fig. 1b). In a hard supercon-ductor, however, this is inhibited by the pinning forces which allow a vortex movement only as long as the driving force P_D is larger than the pinning force P_V. This means that the stationa-ry flux distribution is given by the condition (2.20). Since P_V usually decreases with increasing flux density (see Chapters 3 and 4), the resulting flux distribution will qualitatively look as shown in Fig. 6b. Notice that the boundary between vortex-

16

free and vortex-carrying regions is rather sharp. This is due to the factor dH/dB in Eq. (2.20a) which goes to zero for $B \to 0$ (see Fig. 5), i.e., if dH/dr is finite, dB/dr goes to infinity. Increasing the external field results in a movement of the flux towards the center of the cylinder until the whole cross section is occupied (Fig. 6c). Further increase in H_o increases the flux density more and more (Fig. 6d) until a uniform distribution is reached at the upper critical field H_{c2} where the bulk of the sample goes into the normalconducting state (Fig. 6e).

Reducing H_o from above H_{c2} results in a sign reversal of the flux density gradient (Fig. 6f) because now the driving force points away from the axis, trying to decrease B to the equilibrium value $B_o = B_o(H_o)$. If H_o is reduced below H_{c1} the sample contains "frozen-in" flux (Fig. 6g) which shows up as a large paramagnetic moment similar to that of a permanent magnet. During a further decrease of H_o, the flux distribution $B(r)$ remains unchanged until the external field is reversed and exceeds $- H_{c1}$ (Fig. 6h).

In the second example we will follow the flux changes in a cylinder due to an A.C. field superimposed on a static longitudinal field $H_o > H_{c1}$ (Section 7.3). We will assume that both the frequency $1/T$ and the amplitude h_o of the A.C. field are so small that the flux changes can still be regarded as quasistatic, i.e., we may neglect viscous flux flow forces, etc. At a time $t = t_o$ (Fig. 7a) where the total external field is maximal, we expect a flux distribution similar to Fig. 6d (Fig. 7b). Decreasing the field to the value at $t = t_1$ results in the macroscopic $B(r)$ distribution of Fig. 7c. At $t = T/2$ the total flux ϕ $(- h_o)$ in the sample is a minimum (Fig. 7d). With increasing external field, ϕ is also increased (Fig. 7e) until the cycle is completed at $t = T$ where the maximum flux is reached again (Fig. 7b). We will see in Section 7 that the measurement of the flux changes due to such field cycles provides a convenient method for the determination of pinning forces.

In the preceding examples of the build-up of flux distributions we have assumed that the flux density close to the surface is equal to the equilibrium value $B_o = B_o(H_o)$ given by

17

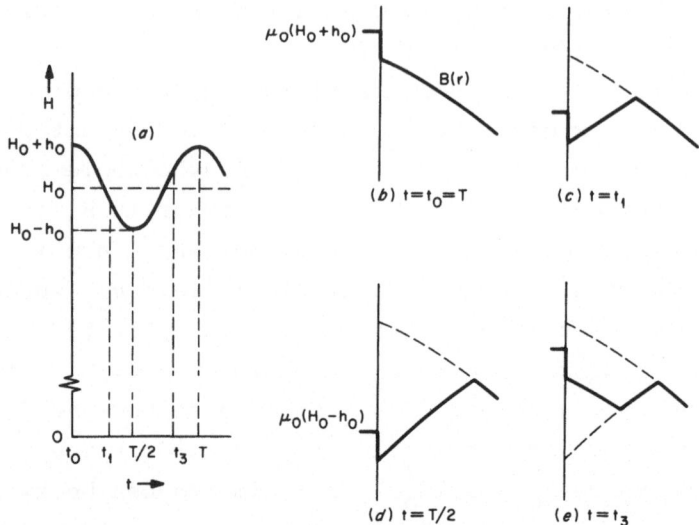

Fig.7 Macroscopic flux density distribution B(r) in a region near the surface of a cylinder which is subject to a small AC-field superimposed on a
large static field H_o. The distributions (b) to (e) correspond to different
times of the field cycle shown in (a).

Eq. (2.5). It should be mentioned that this is not always the
case because of surface currents shielding the bulk of the
sample against small external field changes (ULLMAIER and GAU-
STER 1966). Some of these surface phenomena in hard superconductors and their influence on the flux distribution are discussed in Chapter 7.

2.1.2   Temperature gradients

        The presence of a driving force on flux lines because of
a temperature gradient in a sample can be visualized in a naive
way by considering the temperature dependence of the penetration depth λ. Vortices in the hotter region of a sample repel
each other more strongly due to their larger "diameter" ($\approx \lambda$)
and thus they will tend to move to a region of lower temperature.
This argument considers only magnetic interactions between vortices and neglects the temperature dependence of the maximum
field in the flux line center. Another qualitative explanation
would be that the degree of disorder in the vortex system is
larger at higher temperatures and the entropy can be increased
by flux lines moving towards the colder part of the sample.

In considering driving forces caused by temperature gradients the longitudinal slab geometry used before does not any longer represent the most simple situation. If, for example, a temperature difference is created across the sample in Fig. 1b the equilibrium flux densities are different for the left and right side of the specimen and a flux distribution similar to that in Fig. 4 would arise.

In order to avoid the complication of a combined influence of both a flux density and a temperature gradient, we turn to the flat disk geometry of Fig. 8 which is used in most experiments

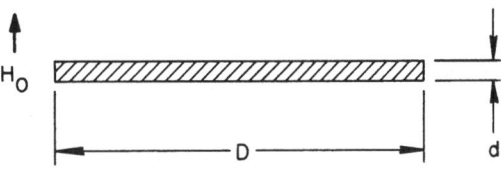

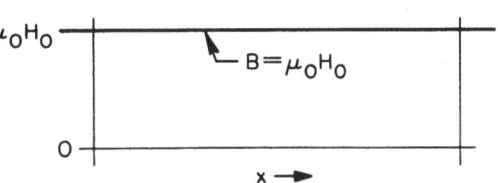

Fig.8 In a flat disk of diameter D much larger than the thickness d, the flux density B is about equal to the external field $\mu_o H_o$.

on thermomagnetic phenomena in superconductors. In the case of a non-zero demagnetization coefficient N, the internal field H and the flux density B are related to the external field H_o by

$$H = H_o - N \left(\frac{B}{\mu_o} - H\right) . \tag{2.27}$$

For the flat disk of Fig. 8, $N = 1 - (\pi d/D) \simeq 1$, and

$$B \simeq \mu_o H_o . \tag{2.28}$$

Therefore in a uniform external field, B is always constant independent of temperature gradients, pinning forces, etc.

An expression for the non-isothermal driving force on a system of straight flux lines was first derived by STEPHEN

19

(1966). He considered the situation where a thermal gradient exists and at the same time a current is passed through the superconductor to prevent flux flow. In mechanical equilibrium the "pressure", $- G_s(T,H) = - F_s + HB$, of the flux line system is constant and with $dF_s = - s^+ dT$ and $B = $ constant we obtain

$$\nabla G_s = 0 = - s^+ \nabla T - B \nabla H .\tag{2.29}$$

Equation (2.29) indicates that the driving force density due to the temperature gradient ∇T is given by

$$P_D = - s^+ \nabla T .\tag{2.30}$$

s^+ is the "transport entropy" of the vortex system per unit volume. s^+ is different from the absolute entropy S of the mixed state and its meaning is discussed in detail by CAMPBELL and EVETTS (1972, p. 227).

Since s^+ is related to the coefficients of the different thermomagnetic effects it can be experimentally determined by thermomagnetic measurements. Many of these effects turn out to be orders of magnitudes larger than in normal metals as was shown by FIORY and SERIN (1966) for the transverse Peltier effect and by OTTER and SOLOMON (1966), and LOWELL (1967) for the Ettinghausen and Nernst effects.

The generalization of Eq. (2.30) is from EVETTS et al. (1968) who found a formally identical expression for the case of curved flux lines. In order to prevent vortex movements the driving force must be balanced by pinning forces. By analogy with the case of flux density gradients we may define a critical state by

$$\underline{P}_D + \underline{P}_v = 0 \quad \text{or} \quad s^+ \nabla T = \underline{P}_v .\tag{2.31}$$

If both flux density and temperature gradients are present the condition for mechanical equilibrium becomes

$$\underline{B} \times \text{curl } \underline{H} + s^+ \nabla T = \underline{P}_v .\tag{2.32}$$

20

Equation (2.32) together with experimental results on s^+ can be used to estimate the relative importance of the two terms in unpinning the flux lattice. Typical isothermal critical current densities J_c = dH/dx for commercial hard superconductors (NbTi) are around 10^9 Am^{-2} at 4.2 K and B = 3T, which corresponds to a driving force density of 3×10^9 Nm^{-3}. The order of magnitude of s^+ can be estimated to be 10^3 Wsecm^{-3}K^{-1} which results in a driving force density of 10^3Nm^{-3} per unit temperature gradient. Therefore, in order to have a depinning effect similar to that of the flux density gradient, ∇T must be around 10^6 Km^{-1}. Such large temperature gradients can never be generated in stationary experiments. However, it might be possible that they do occur locally during the short time of a flux jump (see Chapter 6) and herewith aggravate the effect of the instability.

Very little is known about the effect of accelerations as a cause for macroscopic forces on flux lines. This topic is closely connected to the presence of vibrational modes in the vortex system which is still an open question. One can, however, try to get a feeling about the order of magnitude of the effect of accelerations by using a theoretical expression for the effective mass M'_{eff} of a flux line (see e.g., KIM and STEPHEN 1969)

$$M'_{eff} = \eta\, \tau\, \cos^2 \alpha_H = \frac{\phi_o}{B}\, M''_{eff} \tag{2.33}$$

Here η is the flux flow viscosity, τ the collision time of electrons in the normal state and α_H is the Hall angle ($\alpha_H \approx 0$ in dirty materials). Inserting typical numbers for η, τ, and B we obtain an effective mass M''_{eff} per unit volume of the order of 10^{-2} kg m^{-3}, i.e., even for accelerations as high as 1000 g ($\approx 10^4$ msec^{-2}) the driving force density, i.e. 10^2 Nm^{-3}, will be negligible compared to pinning forces in most hard superconductors. However, inertial forces could be sufficient to trigger flux movements in samples with very low pinning strengths ($P_v \approx 10^4$ Nm^{-3}). Indeed some experimentalists occasionally observed that the magnetization of some almost reversible materials was nearer the equilibrium value if the samples were sha-

ken rigorously before the measurement was performed (e.g., KER-
NOHAN 1965).

2.2 Local Forces

Local forces usually result from interactions between a
flux line and a crystalline defect (pinning). It will be shown
in Chapter 4 that the lattice distortions caused by these local
forces are essential for the transfer of a macroscopic load to
the array of pinning centers. Elasticity theory will be used in
the following to derive expressions for the displacements of
the lattice due to these local forces. For this purpose we will
postpone at this time the discussion of the different pinning
mechanisms (see Chapter 3) and merely assume that the local forces can take the form of point-, line-, or area-forces, depending on the size, shape, and orientation of the pinning centers
with respect to the flux line.
    Local forces are not only caused by crystalline defects
but may also arise from defects in the flux line lattice itself.
Since there are indications that in some cases flux lattice defects can be important for the behavior of hard superconductors,
they will be briefly discussed later in section 2.2.2.

2.2.1 Displacements due to pinning interactions

Because of the strong coupling of flux lines to each other
their displacements resulting from the pinning interactions
(see Chapter 4) are very small, usually much less than the flux
vortex spacings. Elasticity theory can thus be conveniently
used to calculate the displacements u_o of the flux lattice under the influence of a local force K. The elastic properties
of flux lattices may be described by elastic constants C_{ij}
which relate the stresses σ_i to the strains ε_j of the flux
lattice by Hooke's law

$$\sigma_i = C_{ij}\, \varepsilon_j \tag{2.34}$$

We will assume that the z-axis is parallel to the flux lines. This implies that the properties of a hexagonal lattice in the xy-plane are isotropic (HEARMON 1961) if the superconductor itself is isotropic. Since displacements of flux lines parallel to their direction do not affect a two-dimensional array, forces cannot depend on $\varepsilon_{zz} = \varepsilon_3$, i.e. the third column and the third row of the C_{ij} for a hexagonal crystal vanish in our case and we have

$$
\begin{pmatrix} \sigma_{xx} \\ \sigma_{yy} \\ \sigma_{yz} \\ \sigma_{xz} \\ \sigma_{xy} \end{pmatrix} = \begin{pmatrix} C_{11} & C_{12} & 0 & 0 & 0 \\ C_{12} & C_{11} & 0 & 0 & 0 \\ 0 & 0 & C_{44} & 0 & 0 \\ 0 & 0 & 0 & C_{44} & 0 \\ 0 & 0 & 0 & 0 & C_{66} \end{pmatrix} \cdot \begin{pmatrix} \varepsilon_{xx} \\ \varepsilon_{yy} \\ \varepsilon_{yz} \\ \varepsilon_{xy} \\ \varepsilon_{xy} \end{pmatrix} \qquad (2.35)
$$

where $C_{66} = 1/2 \ (C_{11} - C_{12})$. $\qquad (2.36)$

Some of the elastic constants of flux lattices were first calculated by MATRICON (1964) and FETTER et al. (1966). Avoiding the errors made in these early treatments, LABUSCH (1967, 1969b) has recalculated all the elastic constants. They can be expressed in terms of the reversible B(H)-curve of the superconductor, i.e., a measurement of the magnetication curve is equivalent to a measurement of the elastic constants.

The modulus C_L for a deformation that changes only the size of the flux lattice unit cell and not its shape (uniform compression) is given by

$$
C_L = \frac{1}{2} \ (C_{11} + C_{12}) = C_{11} - C_{66} = B^2 \frac{dH}{dB} \qquad (2.37)
$$

Since C_{66} is always much smaller than C_L (see Fig. 9 and Eqs. 2.42 and 2.43) we obtain

$$
C_{11} \approx C_{12} \approx B^2 \frac{dH}{dB} \qquad (2.38)
$$

23

A deformation which tilts a bundle of vortices away from the z-direction while leaving its cross section in the x-y plane unchanged is determined by the constant C_{44} given by

$$C_{44} = B \cdot H \qquad\qquad (2.39)$$

The above formulas for C_{11}, C_{12} and C_{44} are valid within the entire range of $0 \leq B \leq B_{c2} \equiv \mu_o H_{c2}$, because identical expressions are obtained in the limit of well separated vortices as well as in the validity range of the linearized Ginzburg-Landau equation. For the shear modulus C_{66}, however, the low and high field approach give different results. LABUSCH (1967 and 1969b) obtained

$$C_{66} = \frac{\mu_o}{2} \int_o^B B'^2 \frac{d^2 H(B')}{dB'^2} dB' \qquad \text{for } B \ll B_{c2} \qquad (2.40)$$

$$C_{66}' = 0.48 \frac{\mu_o H_c^2 \kappa^2 (2\kappa^2 -1)}{\left[1+\beta_A (2\kappa^2-1)\right]^2} (1 - \frac{B}{B_{c2}})^2 \qquad \text{for } B \lesssim B_{c2} \qquad (2.41)$$

where $\beta_A = 1.16$ for a hexagonal lattice and 1.18 for a square lattice. Typical values of the elastic constants as functions of flux density and temperature are shown for a NbTa alloy in Figs. 9 and 10, respectively.

For high-κ materials and for fields not too close to H_{c1} the elastic constants can be approximated by

$$C_{11} \simeq C_{12} \simeq C_{44} \simeq \mu_o H_{c2}^2 b^2 \qquad\qquad (2.42)$$

$$C_{66}' \simeq \frac{0.13}{\kappa^2} \mu_o H_{c2}^2 (1-b)^2 \qquad\qquad (2.43)$$

where $b \equiv B/B_{c2}$.

24

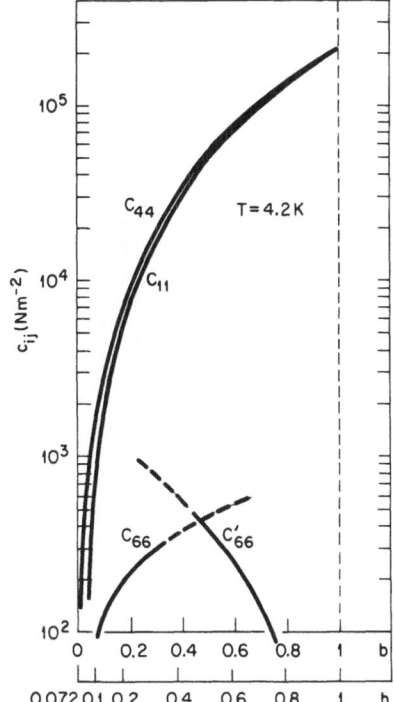

Fig.9 The flux lattice elastic constants for a NbTa alloy T_c = 7.1 K, κ = 3.4 as functions of the reduced flux density b = B/B_{c2}. For the shear modulus both the low field (C_{66}) and the high field (C'_{66}) approximation are shown (ULLMAIER and KERNOHAN 1970).

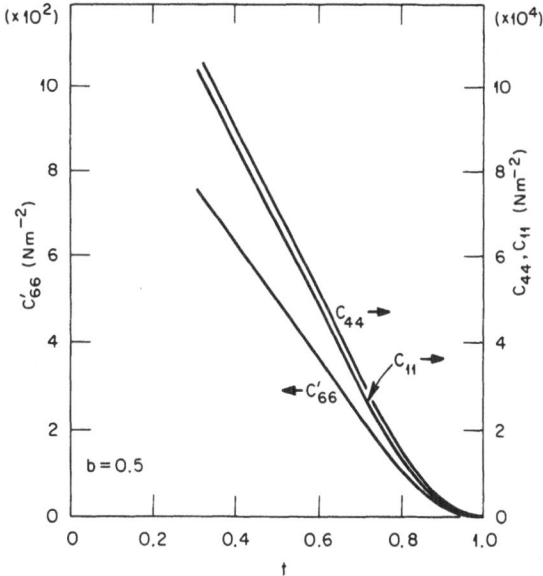

Fig.10 Temperature dependence of the elastic constants for the Nb-Ta alloy of Fig.9 at b = 0.5 (ULLMAIER and KERNOHAN 1970).

25

If the elastic constants of the flux lattice are known then the lattice displacements due to local forces of different kinds can be calculated. The simplest case is that of an area force K/A applied over a plane of area A in the vortex lattice (Fig. 11). If L_x is the distance from the plane A to the plane where the lattice is held stationary, we obtain for the displacement u_o close to the plane A

$$u_o = \int_0^{L_x} \varepsilon_{xx} \, dx = \int_0^{L_x} \frac{\sigma_{xx}}{C_{11}} \, dx = \frac{K}{A} \frac{L_x}{C_{11}} \qquad (2.44)$$

A typical example for the application of this equation is a random distribution of very large particles acting as pinning centers. Although shear and tilt distortions can occur between particles and therefore this situation does not exactly correspond to the ideal picture of Fig. 11, Eq. (2.44) seems to provide a reasonably accurate description for this case (CAMPBELL et al. 1968, see also Chapter 4).

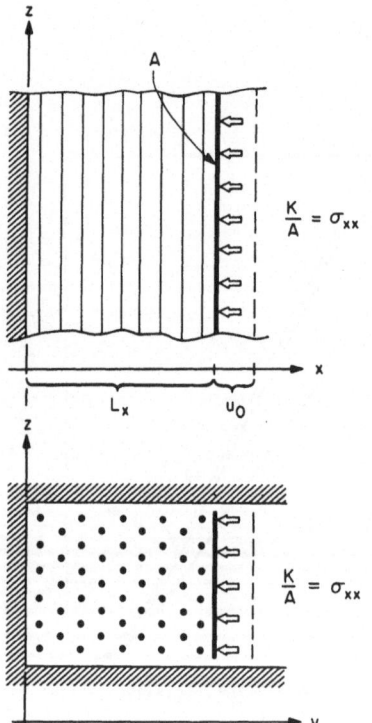

Fig.11 Displacement u_o of a flux line lattice under the influence of an area force K/A. The lattice is assumed to be fixed at the cross-hatched boundaries, i.e. only displacements in the x-direction are possible $(\varepsilon_{yy} = \varepsilon_{yz} = \varepsilon_{xy} = \varepsilon_{xy} = 0)$.

The displacements caused by a line force K/L_z per unit length were first calculated by GOOD and KRAMER (1970). They found that the displacement u_o in the direction of the line force at r_o (an inner cut-off radius introduced to avoid the singularity at $r = 0$, see Fig. 12) is given by

$$u_o = \frac{K}{L_z} \frac{5}{16\pi C_{66}} \ell n \frac{R}{r_o} \qquad (2.45)$$

where R is the distance to the fixed boundary of the flux lattice. r_o is usually set equal to the flux lattice parameter a_o but the choice of r_o (and R) is not critical since it appears only logarithmically in Eq. (2.45).

Classical electrodynamics were used by CAMPBELL and EVETTS (1972, p. 238) to calculate the displacement due to a line force K/L_y acting along a line perpendicular to the vortex direction. For the limit $C_{66} \to 0$ they found

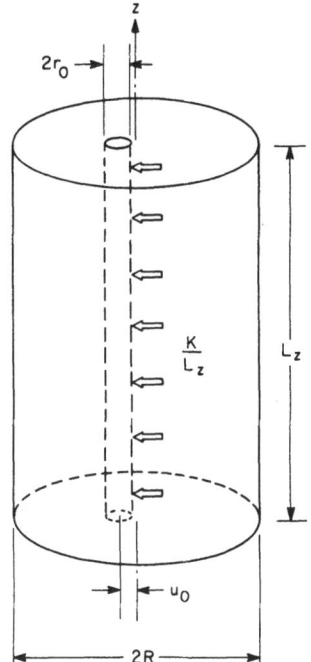

Fig.12 Displacement u_o of a flux line lattice under the influence of a line force K/L_z. R is the radius of the cylindrical boundary where the flux lattice is fixed and r_o is the radius, where the displacement is measured. The flux line direction is parallel to the z-axis.

$$u_o = \frac{K}{Ly} \frac{1}{2\pi C_{11}} \ell n \frac{R}{r_o} \qquad (2.46)$$

Since it was assumed that the vortex arrangement is not fixed at some finite distance above and below the line force, no tilting of vortices will occur. In the more realistic situation of a distribution of line forces with an average distance L_z, the constant C_{44} must appear in the expression for u_o and it is probably more appropriate to consider such line forces as a sequence of point forces.

Examples of line forces are crystal dislocations interacting with flux lines via elastic interactions (see Chapter 3) or forces due to defects in the flux lattice itself (see following section).

For the case of a point force K acting at the origin in the direction of the x-axis the displacement in an isotropic lattice at a distance r_o is given by

$$u_o(r_o) = \frac{K}{4\pi C_e r_o} \qquad (2.47)$$

where C_e is an effective modulus which is a function of the elastic constants C_{ij}. u diverges at the point where the force is applied and this point is of interest for pinning theory. Therefore LABUSCH (1969b) calculated u_o for a force distributed over the area of a unit cell of the flux lattice, but localized to a single value of the z-coordinate (Fig. 13). The result is

$$u_o = K \left(\frac{B}{\phi_o}\right)^{1/2} \frac{1}{4\sqrt{\pi}} \left[(C_{44} C_{66})^{-1/2} + (C_{44} C_{11})^{-1/2} \right] \qquad (2.48)$$

Since C_{66} (and also C_{66}') is always much smaller than C_{11}, the last term in Eq. (2.48) can usually be neglected and the effective modulus C_e defined in Eq. (2.47) is approximately

$$C_e \simeq \left(\frac{\sqrt{3} C_{44} C_{66}}{2\pi}\right)^{1/2} \qquad (2.49)$$

with r_o in Eq. (2.47) set equal to $a_o = (2\phi_o/\sqrt{3} B)^{1/2}$. In deriving Eq. (2.48) it was presumed that there is no restriction

in the free volume for displacement, i.e. Eq. (2.48) will only
be accurate if the distribution of point forces is very dilute.
If the spacing ℓ between pinning points becomes smaller, u_o calculated from Eq. (2.48) will be too large since neighborings
point forces cause a cut-off to the free volume of displacement.
If, on the other hand, ℓ becomes very small, the point forces
may act cooperatively thus leading to u_o values larger than given by Eq. (2.48). In order to examine the influence of the pin
spacing ℓ on a more quantitative basis, KRAMER (1973) introduced
a very useful simple model of a flux line interacting with pins.
In Fig. 14 a flux line is subject to point forces K spaced a

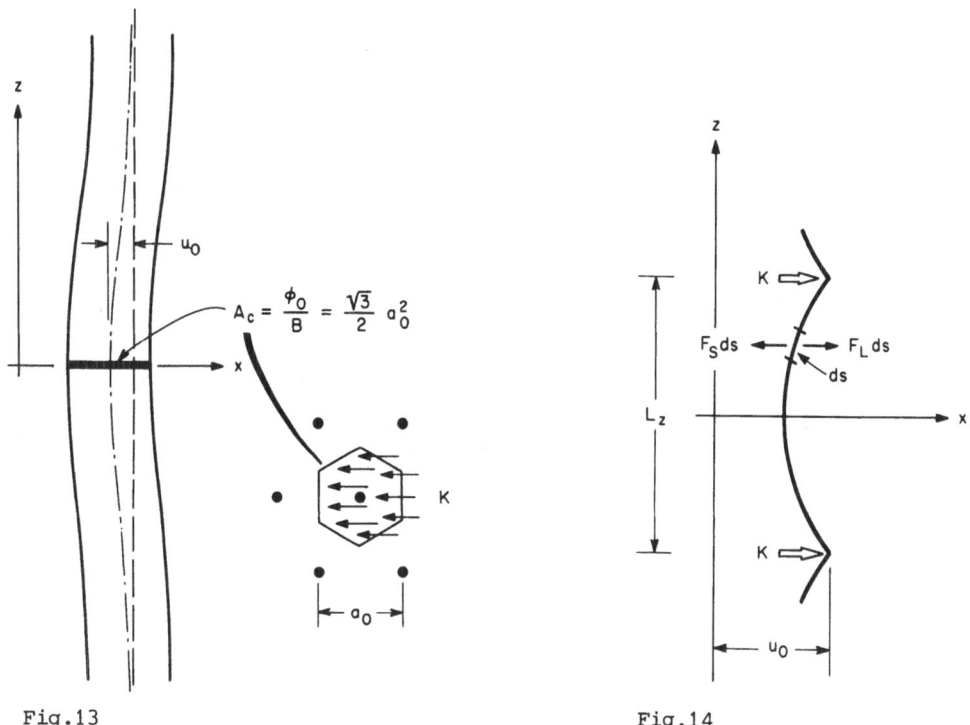

Fig.13   Fig.14

Fig.13 Displacement u_o of a flux lattice under the influence of a "point"
force K perpendicular to the flux line direction z. In order to avoid a di-
vergence of u_o, the force K is assumed to be distributed over the area A_c
of a unit cell of the flux lattice (LABUSCH 1969b).

Fig.14 Model of a flux line, subject to point forces K spaced a distance
L_z apart. Both line tension F_L and shear forces F_s are considered (after
KRAMER 1973).

distance L_z apart. In the region between the pins the equilibrium position of each line element ds is given by the balance of two forces: a force F_L per unit length which is due to the line tension T and a force F_S caused by the surrounding lines of the flux lattice. This leads to the differential equation

$$F_L + F_S = 0 = \frac{C_{44}\phi_o}{B} \frac{d^2x}{dz^2} - C_{66}x \tag{2.50}$$

which determines the shape x(z) of the deformed flux line. In Eq. (2.50) the line tension T_L is expressed in terms of elastic constant C_{44} by the use of Eq. (2.35)

$$\sigma_{xz} = C_{44}\, \epsilon_{xy} = T \frac{B}{\phi_o}\, \epsilon_{xy} \quad ; \quad T = \frac{C_{44}\,\phi_o}{B} \tag{2.51}$$

Neglecting compressional forces, the effect of the neighboring vortices can be approximated by a force due to shearing of the flux line past lattice planes before and behind it. This force may be expressed as

$$F_S = -\, C_{66}x \tag{2.52}$$

The point force K at $z = \pm\, L_z/2$ is given by

$$K = 2\, T\, \frac{dx}{dz}\bigg|_{z\,=\,L_z/2} \tag{2.53}$$

Solving Eq. (2.50) with the boundary condition (2.53) yields for the flux line shape

$$x(z) = K\left(\frac{B}{\phi_o}\right)^{1/2} \frac{1}{2(C_{44}\,C_{66})^{1/2}} \frac{\cosh\left[z\,(C_{66}B/C_{44}\phi_o)^{1/2}\right]}{\sinh\left[L_z/2\,(C_{66}B/C_{44}\,\phi_o)^{1/2}\right]} \tag{2.54}$$

Defining a critical length $\ell^+ \equiv (C_{44}\,\phi_o/C_{66}\, B)^{1/2}$ we obtain for the displacement u_o at the point of application of K

30

$$u_o = x \left(\pm \frac{Lz}{2} \right) = K \left(\frac{B}{\phi_o} \right)^{1/2} \frac{1}{2(C_{44} \, C_{66})^{1/2}} \coth \left(\frac{L_z}{2\ell^+} \right) \qquad (2.55)$$

For a very dilute array of point forces Eq. (2.55) reduces to

$$u_o = K \left(\frac{B}{\phi_o} \right)^{1/2} \frac{1}{2(C_{44} \, C_{66})^{1/2}} \qquad (L_z \gg \ell^+) \qquad (2.56)$$

which is of the same form as the more exact Eq. (2.48) for a single point force. On the other hand, for a concentrated array of pins, Eq. (2.55) yields

$$u_o = \frac{K}{L_z} \frac{1}{2 \, C_{66}} \qquad (L_z \ll \ell^+) \qquad (2.57)$$

which, apart from a numerical factor of order unity has the same form as Eq. (2.45) for a line force. This shows that, in spite of the simplicity of Kramer's model, Eq. (2.55) seems to provide a useful connection between the two limits of point-like and line-like forces, respectively.

The different expressions for u_o given above are valid for the case of small pinning forces, i.e. the displacements are usually much smaller than the vortex spacing a_o. The response of the flux line arrangement is then entirely determined by the rigidity of the flux lattice and this case is therefore called "lattice approximation". For the case of point forces LABUSCH (1969a) has derived a validity criterion for the lattice approximation. He finds that Eq. (2.48) is correct if

$$\alpha_L / 4\pi \, C_{66} \ll 1 \qquad (2.58)$$

where $\alpha_L = \langle \nabla \nabla U \rangle$ and U is the interaction potential per unit length between the vortices and the crystal lattice. Thus the parameter $\alpha_L / 4\pi \, C_{66}$ corresponds to the ratio between the mean strength of the pinning interaction and the interaction of flux lines with each other. Inserting numerical values for $\nabla \nabla U$, estimated for different pinning mechanisms into Eq. (2.58) shows that the lattice approximation applies in all practical cases except for fields very close to H_{c1}. This result is supported by low amplitude A.C. measurements (CAMPBELL and EVETTS 1972, p. 361) and neutron diffraction experiments (LIPPMANN et al. 1975, see Section 7.5).

For very low flux densities, C_{66} becomes very small and we expect a fluid-like response of the flux line arrangement. In this "fluid approximation" the displacement u_0 caused by a point force K is given by (LABUSCH 1969a)

$$u_0 = K \left(\frac{B}{\phi_0}\right)^{1/2} \frac{1}{4\sqrt{\pi}} \left(\frac{1}{(C_{11}C_{44})^{1/2}} + \frac{\sqrt{\pi}}{(\alpha_L C_{44})^{1/2}} \right) \qquad (2.59)$$

for $\dfrac{\alpha_L}{4\pi\, C_{66}} \gg 1$ and $\dfrac{\alpha_L}{4\pi\, C_{11}} \ll 1$.

A fluid-like structure has been observed in Pb-Tl for $B < 2.5$ mT by ESSMANN and TRÄUBLE (1969).

For even smaller flux densities the interaction between vortices is negligible and the response depends only on the line tension $T = C_{44}\, \phi_0/B$. In this "single vortex approximation" u_0 is given by

$$u_0 = K \left(\frac{B}{\phi_0}\right)^{1/2} \frac{1}{2(\alpha_L C_{44})^{1/2}} \qquad (2.60)$$

for $\dfrac{\alpha_L}{4\pi\, C_{11}} \gg 1$

It should be mentioned that the criteria (2.58) to (2.60) are valid only for point forces acting on an ideal lattice. For line forces and/or the presence of flux lattice defects (see next Section) the changeover from the lattice approximation to the fluid approximation may occur at considerably smaller α_L values as given by Eq. (2.59). Recent numerical calculations (SCHMUCKER 1974, ESSMANN and SCHMUCKER 1974) indicate that a low flux densities the flux lattice dislocations can be moved so easily that treating the vortex lattice as a fluid corresponds to a good approximation for many purposes.

2.2.2 Flux line lattice defects

The occurrence of dislocations in the flux line lattice was first predicted by LABUSCH (1966a). Shortly afterwards, ESSMANN and TRÄUBLE (1967) developed a decoration technique which permits a direct observation of the positions of individual flux lines piercing the surface of a type II superconductor (see Chapter 7). Their first experiments showed that dislocations are indeed present in the flux line lattice, sometimes together with other defects such as stacking faults, vacancies, interstitials and grain boundaries. In this context we will discuss these flux lattice defects only with respect to their possible importance for hard superconductors. For a detailed description the reader is referred to the original papers by TRÄUBLE and ESSMANN (1968a,b) and ESSMANN and TRÄUBLE (1969) or to the summary given by SEEGER (1968).

Vacancies, interstitials and their agglomerates (Fig. 15) are observed in a very narrow range of low flux densities only,

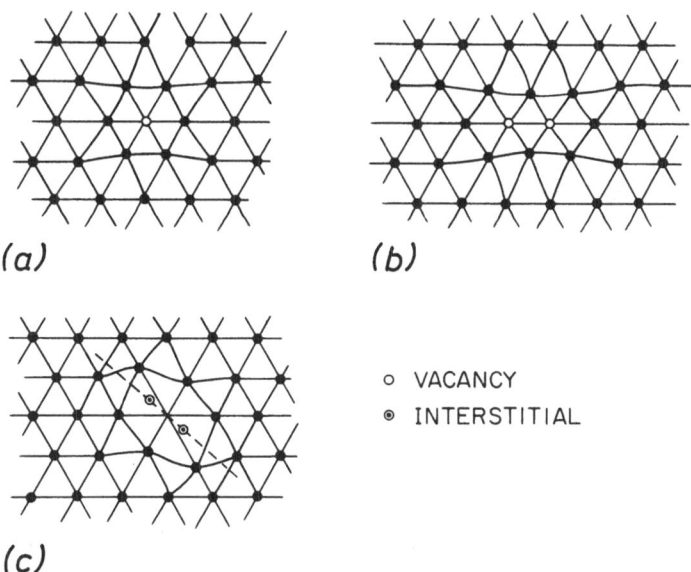

(a)

(b)

(c)

○ VACANCY

◉ INTERSTITIAL

Fig.15 Examples of line defects in the flux line lattice: (a) vacancy, (b) divacancy, and (c) interstitial in dumbell configuration (schematic). Because of the greater distortion of the surrounding lattice, interstitial lines are less frequently observed than vacancy type defect lines (after TRÄUBLE and ESSMANN 1968a).

e.g. in Pb-6.3% In at T = 1.2 K such line defects were found for 3 mT<B<6 mT (below 3 mT the vortex lattice dissolves into a fluid structure and the term "lattice defect" becomes meaningless). The line defect concentration found in this field range was around 10^{-4} per flux line which is orders of magnitude too large for the defects to be in thermodynamic equilibrium. Therefore line defects must be produced during the formation of the flux lattice, similar to "grown in" defects in crystal lattices (dislocations, stacking faults, etc.).

Theoretical considerations based on the London model of a flux line were carried out by BRANDT (1969a) and HILL et al. (1969). For fields near H_{c2} BRANDT (1969b) gave a microscopic description of a special type of vacancy using the Ginzburg-Landau theory. Both approaches and the few experimental results suggest that except when the field is very close to H_{c1} the influence of line defects on the macroscopic behavior of the vortex lattice will be very small.

In contrast to this, there are indications that dislocations (Figs. 16 and 17) can play a substantial role in hard superconductors. LABUSCH (1966a) previously pointed out that edge dislo-

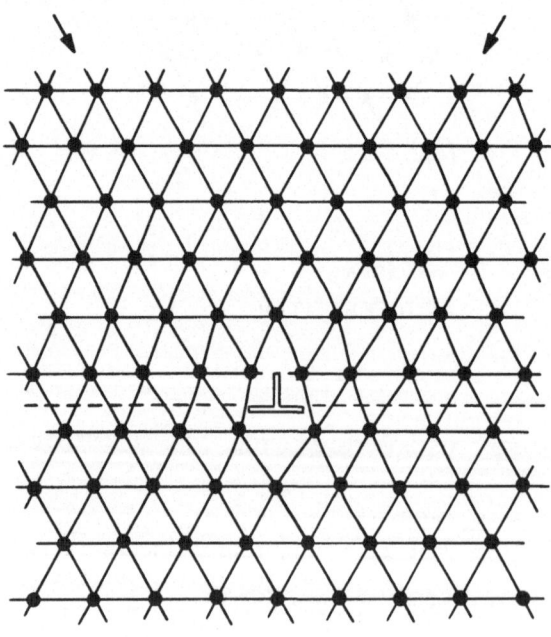

Fig.16 Edge dislocation in a hexagonal flux line lattice. The two extra planes which end at the dislocation are marked by arrows. The dotted line represents the slip plane.

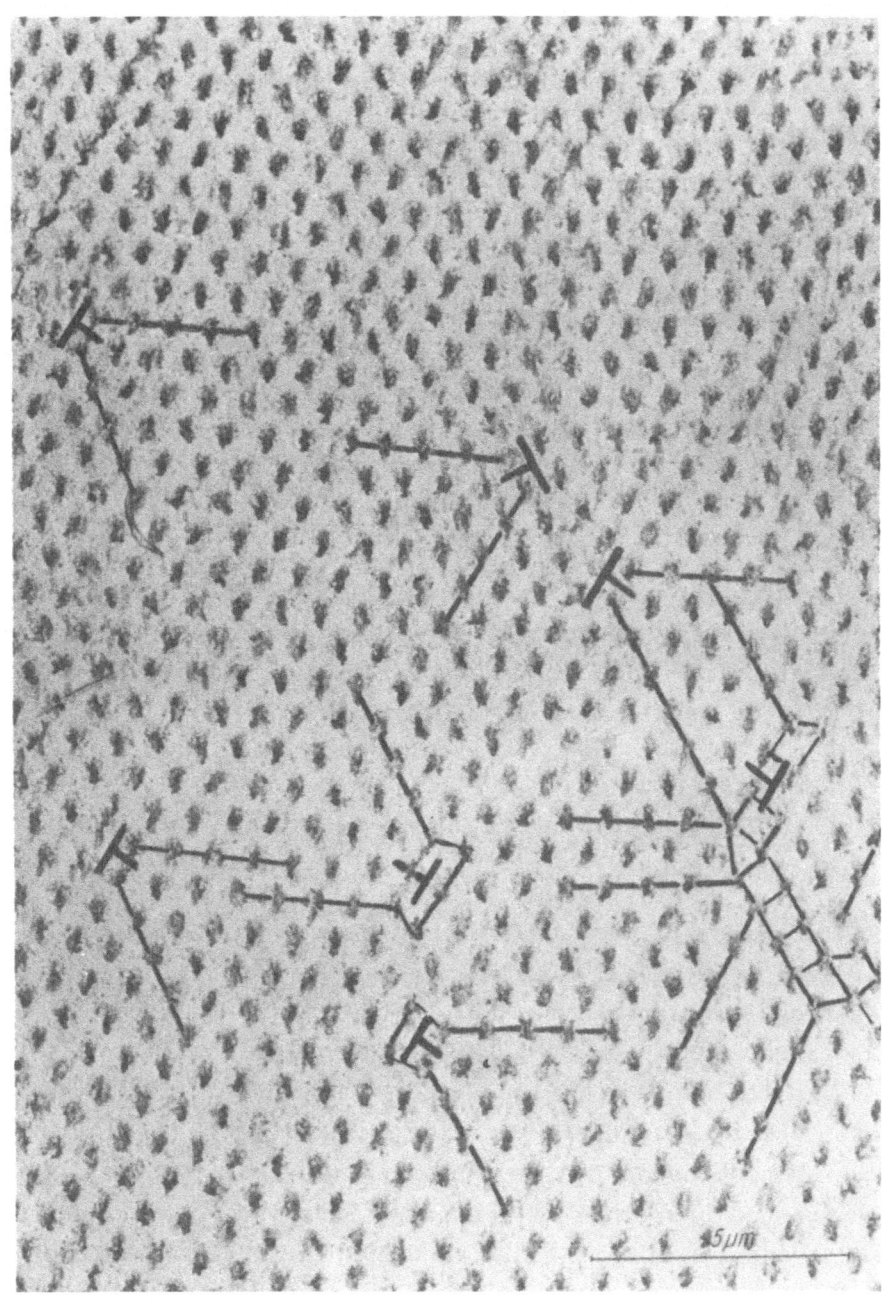

Fig.17 Micrograph of a replica showing a flux line arrangement containing
a high density of dislocations. The dotted lines mark the extra planes
ending at the dislocations. on the lower right side a partial dislocation
connected to a stackung-fault (square lattice) can be seen (TRÄUBLE and
ESSMANN 1968a; Pb-6.3 at. % In sample in remanent state, B ≃ 35 mT, T = 1.2 K).

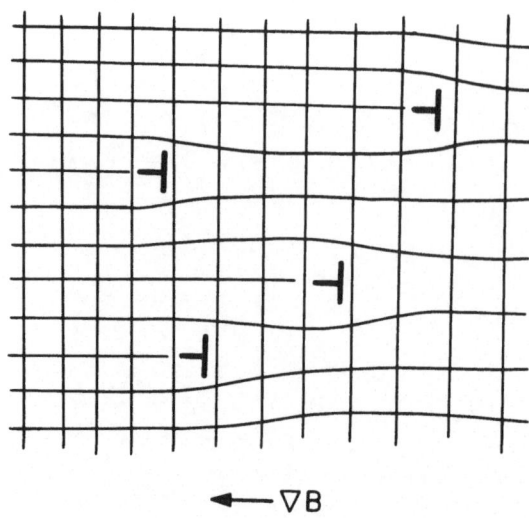

$\longleftarrow \nabla B$

Fig.18 Flux density gradient maintained by edge dislocations in the flux line lattice. For the sake of simplicity the case of a square lattice is shown in this schematic diagram.

cations might be important for the formation of a flux density gradient. A simple geometrical consideration (see Fig. 18) shows that the flux line dislocation density ρ_{FD} necessary to produce a flux density gradient dB/dx is given by

$$\rho_{FD} = \frac{k}{(\phi_o B)^{1/2}} \frac{dB}{dx} \tag{2.61}$$

k is a constant not very different from 0.5 which depends on the lattice type and the orientation of the slip planes of the dislocations with respect to ∇B. By comparing observed ρ_{FD} and dB/dx values ESSMANN and TRÄUBLE (1969) were able to show that Eq. (2.61) is indeed fulfilled within the range of low flux densities accessible to the decoration techniques, i.e. the dislocations alone account for the observed flux density gradient. They suggested that this is generally true, but their conclusion was questioned by CAMPBELL and EVETTS (1972, p. 243) for the following reasons. At a given flux density gradient the dislocation concentration ρ_{FD}/n given by

$$\frac{\rho_{FD}}{n} = \rho_{FD} \frac{\phi_o}{B} = \frac{k' \, \phi_o^{1/2}}{B^{3/2}} \frac{dB}{dx} \tag{2.62}$$

36

decreases rapidly with increasing B; e.g. for $J_c = 2 \times 10^8$ Am^{-2}, $\rho_{FD}/n \simeq 2 \times 10^{-2}$ for B = 5 mT but $\rho_{FD}/n \simeq 5 \times 10^{-7}$ for B = 5 T. The latter concentration corresponds to a mean distance between dislocations of about 30 μm. Since this is certainly much larger than the mean distance between pinning centers the lattice must be distorted on a much finer scale, independent of the flux line dislocations, in order to transfer a macroscopic force to the array of pinning centers (see Chapter 4). Campbell and Evetts therefore suggest that both dislocations and other distortions combine together to give the required flux density gradient. At very low flux densities the effect of dislocations dominates whereas at medium and high flux densities their influence will be small. Recent neutron diffraction experiments which measure the deviations of the flux line directions from the external field direction (LIPPMANN et al. 1975) seem to confirm this view.

Besides their importance in the formation of flux density gradients, there are some indications that flux lattice dislocations can affect the maximum static pinning force. This was shown in an experiment by CAMPBELL and EVETTS (1972) in which they altered the flux lattice defect structure by generating arrays of edge dislocations using miniature coils placed above and below a plate-like specimen (Fig. 19a). It was found that the critical current density J_c varied by a factor of at least three depending on the way in which the flux line arrangement was created. In particular, the decrease of J_c ("degradation") was roughly proportional to the dislocation density produced before the voltage-current characteristic was measured (Fig. 19b). At present there is no unambiguous interpretation of these interesting results, however, they clearly show that in some materials the pinning force density P_V is not a fixed parameter but can vary appreciably with the magnetic history of the sample.

CHANG et al. (1969) were the first who pointed to similarities between flux flow voltage-current characteristics and strain-stress curves in mechanical tests (Fig. 20). They interpreted their results on samples showing a peak in the critical current density in terms of yield and dislocation motion. They further suggest that in the region of the peak effect (see Section 4.3) the macroscopic pinning force density should be described entire-

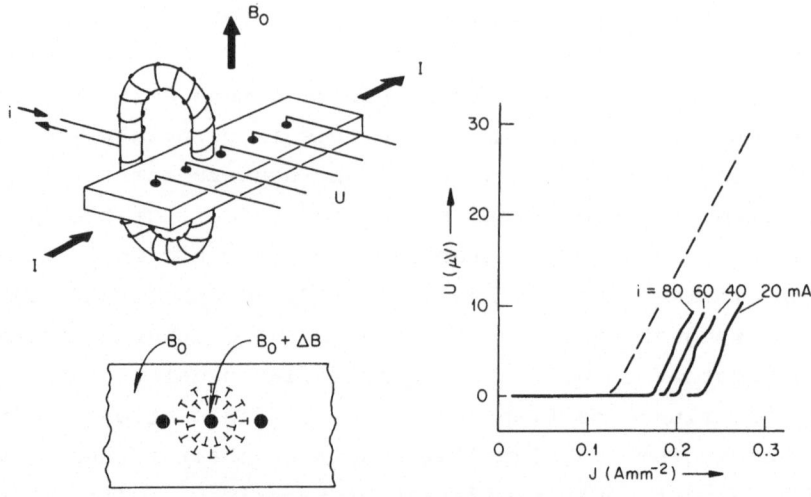

Fig.19 Experiment designed to investigate the influence of vortex lattice defects on pinning parameters. A split ring solenoid locally generates an additional field ΔB. This cause rings of edge dislocation in regions where the flux density changes from B_o to $B_o + \Delta B$ which is schematically shown at the lower left of the figure. The diagram at the right shows the dependence of the critical current density on the dislocation density which is pro- portional to the solenoid current i. The dashed line represents the "fully degraded state" which is reached after cycling the sample fully into the flux flow state (after CAMPBELL and EVETTS 1972, p. 400).

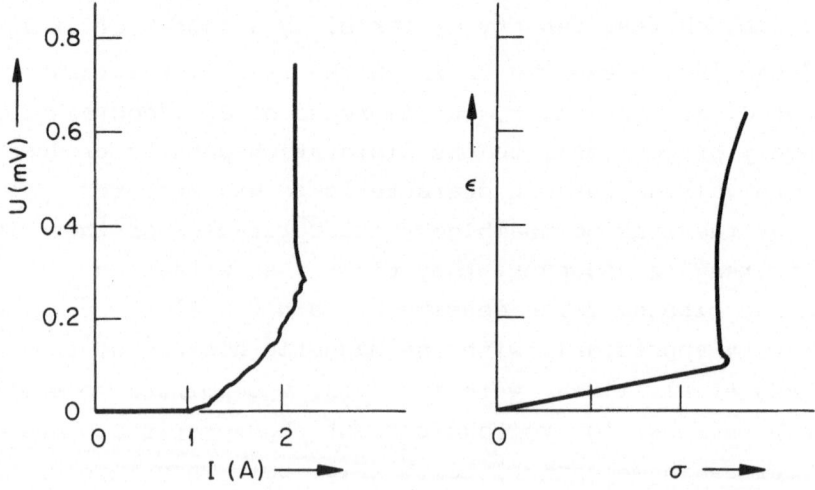

Fig.20 (a) Voltage - current characteristic of a hard superconductor in the flux flow state. Note the similarity with a stress-strain curve shown in (b) for a material exhibiting a sharp yield point (after CHANG et al. 1969).

ly by pinning of flux line dislocations rather than pinning of flux lines themselves.

The influence of the perfection of the vortex arrangement on the flux flow behavior was also studied by GOOD and KRAMER (1971) on a niobium crystal containing weak pinning centers. They measured the current I_F required to maintain a certain constant voltage (i.e. flux flow velocity) across the sample. This procedure is analogous to a mechanical test where a measurement is made of the stress necessary to produce a constant strain rate. Observing the time dependence of the current I_F, Good and Kramer found recovery and yield phenomena quite analogous to those found in mechanical tests. They explain their results by assuming that the flux flow voltage is caused by the motion of dislocation dipoles which are generated by dislocation sources. An expression equivalent to the OROWAN (1940) equation of dislocation dynamics was derived by KRAMER (1970). In order to obtain agreement between the experimental and theoretical forces to operate the dislocation sources, he had to assume an exceptionally large scale of the pinning structure and the diameter of the source. It therefore still seems to be an open question as to the way these sources form and operate (see also CAMPBELL and EVETTS 1972, p. 407). Other discussions of the dynamics of flux lattice defects are due to LABUSCH (1966a), BRANDT (1969c) and ESSMANN and SCHMUCKER (1974).

There are no experimental results related to the influence of grain boundaries on pinning. Both low and high angle grain boundaries in the vortex lattice have been found in materials investigated by the decoration technique (ESSMANN and TRÄUBLE 1969). One possible reason for the formation of flux lattice grains in polycrystalline samples is the coupling between the flux line lattice and the crystal lattice caused by anisotropies of superconducting parameters in real metals (OBST 1971, SCHELTEN et al. 1971, ULLMAIER et al. 1973). Simulation experiments empolying the floating magnet analogue by ROSE-INNES and STANGHAM (1969) indicate that grain boundaries are sinks for other defects (MELVILLE and TAYLOR 1970). Thus in materials with small grains the effect of grain boundaries on the flux flow behavior could be important.

The few experiments mentioned above all employed samples with very weak pinning. There are no investigations on the influence of flux line defects in materials with strong pinning and at high flux densities. At present it is therefore not possible to draw any general conclusions about the importance of vortex lattice defects for the pinning process. There are, however, many indications that their influence will be much less pronounced than that of crystalline defects on the mechanical hardness of materials. The main reason for this difference is probably the following. In contrast to the case of crystal lattices, flux line lattices are pinned by interactions between pinning centers and vortex lattice defects as well as vortices themselves. Theoretical calculations of the maximum interaction between a flux line and a pinning center (see Chapter 3) yield values similar to those obtained for the force between a vortex lattice defect and a pinning center (KUSAYANAGI and YAMAFUJI 1969, CAMPBELL and EVETTS 1972, p. 399). Since the concentration of defects is relatively low (see Eq. 2.62) their contribution to the total pinning force will be small, at least for medium and high flux densities. The suggestion of CHANG et al. (1969) that the total pinning force is determined wholly by pinning of flux line dislocations therefore seems to be very unlikely to hold in the general case.

The preceding discussion shows that more experiments concerning the morphology of flux lattices are required before any valid general conclusions can be drawn. Besides the decoration technique, neutron diffraction methods have been applied to this problem recently. Although until the present time only fairly reversible samples have been investigated (SCHELTEN et al. 1971, 1974; THOREL et al. 1973) it is hoped that the methods can also be used for hard superconductors (LIPPMANN et al. 1975). Used simultaneously with critical current measurements they should be capable of revealing possible correlations between macroscopic pinning parameters and the degree of perfection of the flux line lattice.

3. Pinning Mechanisms

It is now well established that the high supercurrent carrying capacity of hard superconductors is a consequence of the pinning of flux lines by inhomogeneities in the material ("pinning centers"). Such local changes of superconducting properties arise from the presence of structural defects and result in position-dependence of the free energy of the flux lines. The strength of a pinning center can be characterized by a pinning potential U which is the difference between the free energy of a flux line in the pinning center and in the surrounding homogeneous material ("matrix"). The corresponding interaction force is then given by $\underline{K} = \nabla U$.

A number of different pinning interactions has been proposed by several authors but for a given defect structure only rough estimates of the relative importance of these mechanisms can be made at present. For such estimates it is practical to divide the field range between H_{c1} and H_{c2} into the same two regions as in the calculation of equilibrium properties of the mixed state (Chapter 8). In the low field region the free energy of a quasi-isolated flux line is split into contributions from the vortex core and from the surrounding magnetic fields and currents. In the high field region, the flux lattice is described by the Ginzburg-Landau theory. In the following I shall discuss some basic pinning mechanisms and give approximate expressions for their interaction strength as a function of flux density, temperature, superconducting matrix properties, and defect-specific parameters.

3.1 Pinning of Quasi-Isolated Flux Lines

3.1.1 Core interactions

At the center of a flux line the superconducting order parameter goes to zero as e.g. shown in Fig. A3. In order to creat
this normal conducting core of volume V_c, the condensation energy $V_c \mu_o H_c^2/2$ must be supplied. A part of this energy can be conserved if the flux line core passes through a region where the
order parameter is already zero (e.g. a void or a normal conducting precipitate). If the actual core structure is approximated
by a normal conducting cylinder of radius ξ, the maximum pinning
potential U_o can easily be estimated for pinning sites of different size and shape.

For a small spherical void or insulating inclusion of diameter L much smaller than the coherence length ξ, U_o is given
by

$$U_o = \frac{\mu_o H_c^2}{2} \frac{4\pi}{3} \left(\frac{L}{2}\right)^3 \qquad (L \gg \xi) \qquad (3.1)$$

Since the range, over which the order parameter changes, is given by the coherence length ξ, the maximum interaction force K_o
will approximately be equal to U_o/ξ. Using Eq. (8.23) to replace
H_c and ξ in terms of H_{c2} and κ we obtain from Eq. (3.1)

$$K_o = \frac{(2\pi \mu_o)^{3/2}}{48 \, \phi_o^{1/2}} \frac{H_{c2}^{5/2}}{\kappa^2} L^3 \qquad (L \ll \xi) \qquad (3.2)$$

For a void of diameter L = 20 Å in a superconducting matrix[+]
with B_{c2} = 12 T, κ = 50, and ξ = 50 Å at 4.2 K, Eq. (3.2) yields
a core pinning force of about 10^{-14} N. For small normal conducting precipitates Eq. (3.2) probably overestimates K_o since Cooper pairs can leak from the superconducting matrix into the nor-

---

[+] In order to get a feeling about the relative strengts of the different
pinning mechanisms, this "standard" matrix will be used throughout this
chapter. Its values for B_{c2}, κ, and ξ are approximately equal to those of
the commercial alloy $Nb_{50}Ti_{50}$.

mal zone ("proximity effect", see e.g. DEUTSCHER and DE GENNES 1969) and thus diminish the energy difference U_o. An even larger reduction will occur for a superconducting precipitate with thermodynamic critical field $H_c' \neq H_c$. In this case U_o will be given by the difference of the condensation energy in both phases.

For voids larger than the core region the maximum pinning energy per unit length is

$$U_o = \frac{\mu_o H_c^2}{2} \pi \xi^2 \tag{3.3}$$

and the maximum interaction force K_o will depend on the shape and orientation of the void. If there is a sharp void surface of length L_z parallel to the vortices, the force per unit length is

$$\frac{K_o}{L_z} \approx \frac{U_o}{\xi} = \frac{(2\pi \phi_o \mu_o)^{1/2}}{8} \frac{H_{c2}^{3/2}}{\kappa^2} \tag{3.4}$$

Inserting the same matrix parameters as before $(B_{c2} = 12$ T, $\kappa = 50)$ into Eq. (3.4) leads to $K_o/L_z = 2 \cdot 10^{-4}$ Nm^{-1}.

On the other hand, if the particle is a sphere of diameter $L > \xi$, U_o should be divided by $L/2$, the distance over which U changes from zero to its maximum value U_o. We then obtain for K_o acting over a length $L_z \approx L/2$

$$K_o \approx \frac{2U_o}{L} \frac{L}{2} = \frac{\phi_o}{8} \frac{H_{c2}}{\kappa^2} \tag{3.5}$$

In deriving Eqs. (3.1) to (3.5) we have presumed that the flux line structure is not changed by the presence of the pinning center. Whereas this assumption is reasonable for small particles it will hardly be valid for defect dimensions comparable to the characteristic lengths ξ and λ of the flux lattice. Equations (3.4) and (3.5) should therefore be regarded as order of magnitude estimates only. Otherwise, similar arguments as before apply if the void is replaced by a normal conducting or superconducting precipitate or by grains of different superconducting properties in polycrystalline anisotropic materials (see Chapter 5).

3.1.2 Magnetic interactions

In addition to the condensation energy term in the core region, the free energy of a flux line is determined by the sum of the magnetic field energy and the kinetic energy of the vortex currents (see Eq. 8.9). Inhomogeneities which change the field and current distribution of a flux line will therefore introduce a position dependence of its free energy and give rise to the so-called magnetic pinning interaction. A simple example for such an interaction is the effect of a smooth plane boundary between an insulator and a superconductor which is parallel to the vortices and perpendicular to the driving force. A flux line approaching such a boundary will be subject to two forces. Firstly, there is a repulsive force caused by the surface current I_s which corresponds to the flux density difference $M_{rev} = \mu_o H - B$ (see Eq. 2.3 and Fig. 1). We may assume that the surface current density j_s varies as

$$j_s = \frac{\mu_o H - B}{\mu_o \lambda} e^{-x/\lambda} = \frac{M_{rev}}{\mu_o \lambda} e^{-x/\lambda} \qquad (3.6)$$

with the distance x from the surface (see Fig. 21a). This surface current exerts a Lorentz force per unit length of

$$\phi_o j_s = \frac{\phi_o M_{rev}}{\mu_o \lambda} e^{-x/\lambda} \qquad (3.7)$$

on the flux line. Secondly, there is an attractive force which is due to the distortion of the flux line structure in the vicinity of the plane. This attraction can be described as the force caused by an image vortex of opposite direction (BEAN and LIVINGSTON 1964) and will vary as $D^+ e^{-2x/\lambda}$ (see Fig. 21b). In thermodynamic equilibrium, the total work done by moving a flux line from the surface (x = 0) to x = ∞ must vanish. This condition was used by CAMPBELL and EVETTS (1972, p. 340) to find

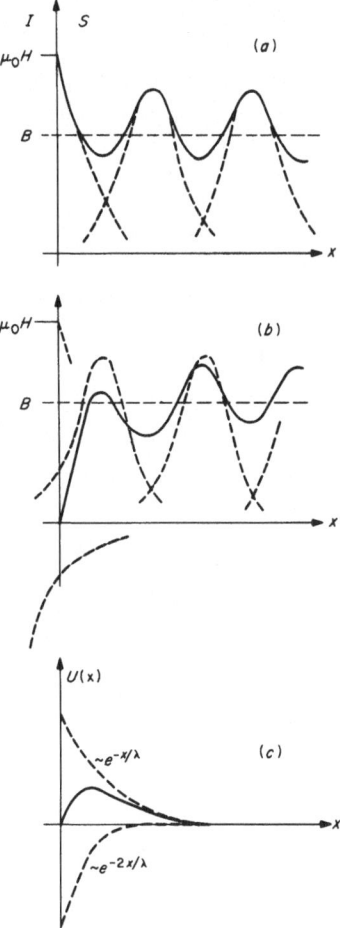

Fig.21 (a) Undistorted microscopic magnetic field distribution h(r) in a row of
flux lines perpendicular to a boundary between an insulator I and a superconductor
S (schematic). The surface current
I_s = H - B/μ_o repels flux lines from the
boundary. In (b) the distortion of h(r)
near the boundary is shown as the sum of
the original distribution and the negative fields from an image vortex. In (c)
the potential energy U of a flux line
resulting from (a) and (b) is shown as a
function of the distance x from the
boundary.

$$D^+ = \frac{2\phi_o M_{rev}}{\mu_o \lambda} \tag{3.8}$$

from
$$\int_0^\infty (\frac{\phi_o M_{rev}}{\mu_o \lambda}) e^{-x/\lambda} - D^+ e^{-2x/\lambda}) \, dx = 0$$

This relation leads to a maximum interaction force per unit
length at x = 0

$$\frac{K_o}{L_z} = - \frac{2\phi_o M_{rev}}{\mu_o \lambda} + \frac{\phi_o M_{rev}}{\mu_o \lambda} = - \frac{\phi_o M_{rev}}{\mu_o \lambda} \tag{3.9}$$

45

For low flux densities ($M_{rev} \simeq \mu_o H_{c1}$) the relative strengths of the core interaction and the magnetic interaction of a plane boundary can be compared by dividing Eq. (3.4) by Eq. (3.9). Replacing H_{c1} by H_{c2} and κ (Eqs. 8.16 and 8.23) one finds

$$\frac{K_o \; core}{K_o \; mag} = \frac{\kappa}{8} \qquad (b \to 0) \qquad (3.10)$$

The major reason for the core interaction to become dominant for high κ-values is the different interaction distance d, i.e. for core pinning $d = \xi$ decreases with increasing κ (see Table A1) whereas for the magnetic interaction $d = \lambda$ increases with κ. For our high-κ "standard" matrix (see footnote on p. 42) we therefore find a magnetic force of $K_o/L_z = 3 \cdot 10^{-5}$ Nm^{-1} which is weaker than the core interaction calculated from Eq. (3.4). Approximating $M_{rev} = B - \mu_o H$ by Eq. (8.25) and using Eq. (8.23) to express $\lambda = \kappa \cdot \xi$ in terms of κ and H_{c2}, Eq. (3.9) may be rewritten as

$$\frac{K_o}{L_z} = \frac{(2\pi \; \phi_o \; \mu_o)^{1/2}}{2} \; \frac{H_{c2}^{3/2}}{\kappa^3} \; (1 - b) \qquad (3.11)$$

Similar considerations as the above will be valid for a boundary between two superconductors with different κ or H_{c2} values. In this case, M_{rev} in Eq. (3.9) should be replaced by the difference ΔM_{rev} of the magnetizations of the two regions.

Examples of pinning centers effective by way of magnetic interaction are large voids or insulating inclusions, large normal conducting precipitates (CAMPBELL et al. 1968, COOTE et al. 1972), and dislocation cell structures with a Ginzburg-Landau parameter difference $\Delta\kappa$ between cells and cell walls (NARLIKAR and DEW-HUGHES 1969). In deriving Eq. (3.9) it was necessary that the regions on either side of the pinning boundary be large enough to define macroscopic flux densities B within them. However, a magnetic interaction will also occur with somewhat smaller particles ($L > \lambda$) and can be estimated in the following way. If a flux line passes through a normal conducting region, not only the core energy but also the kinetic energy of the vortex currents will be conserved. Since the kinetic energy term in Eq. (8.10) is roughly half the line energy E_L, the maximum interaction potential derived from Eq. (8.11) is

$$U_o \simeq \frac{E_L}{2} = \frac{1}{8\pi \, \mu_o} (\frac{\phi_o}{\lambda})^2 \, \ell n \, \kappa \qquad (3.12)$$

which gives rise to a force per unit length of

$$\frac{K_o}{L_z} \simeq \frac{U_o}{\lambda} = \frac{(2\pi \, \phi_o \, \mu_o)^{1/2}}{4} \frac{H_{c2}^{3/2}}{\kappa^3} \, \ell n \, \kappa \qquad (3.13)$$

Here λ has again been replaced by H_{c2} and κ. Apart from the slowly varying factor $2/\ell n\kappa$, Eq. (3.13) is identical to Eq. (3.10) which seems to indicate that the different approaches leading to these relations are equivalent.

Very small particles ($L \ll \lambda$) are not expected to interact magnetically with the flux lines. Ferromagnetic inclusions are exceptions to this general rule. These were investigated by ALDEN and LIVINGSTON (1966). It was found that the pinning force depends on the relative direction of the vortices and the magnetization of the particles.

Finally it should be mentioned that the field and temperature dependence of the core interaction and the magnetic interaction are roughly the same (compare Eqs. 3.4 and 3.13). This makes it very difficult to decide from macroscopic pinning force measurements which of the two mechanisms is dominant in a given experimental situation.

3.1.3 Elastic interactions

The origin of elastic interactions lies in the fact that the specific volume and the elastic compliances of a metal in the superconducting state are slightly larger than in the normal state. Typical fractional changes are of the order of $\varepsilon_{vo} \simeq 10^{-7}$ for the volume dilatation (BRÄNDLI et al. 1968) and $\Delta S/S \simeq 10^{-4}$ for the elastic compliance change (ALERS and WALDORF

1961). For a flux line this means that the metal lattice in the core will be slightly denser and stiffer than in the superconducting region around it. This leads to two interaction mechanisms, one linear in stress (first order or "parelastic" interaction), the other quadratic in stress (second order or "dielastic" interaction).

The parelastic interaction results from the strain fields around the vortex core interacting with the stress field of a defect. A first estimate of the interaction force caused by this effect was given by KRAMER and BAUER (1967). They made the reasonable assumption that the dilatation ε_v varies linearly with the order parameter $|\psi|^2$ in the vortex core. Since $|\psi|^2 = |\psi_0|^2 \left[1-\exp(-r^2/\xi^2)\right]$ we therefore have

$$\varepsilon_v = \varepsilon_{vo} \left(1 - e^{-r^2/\xi^2}\right) \tag{3.14}$$

ε_{vo} is the dilatation of the Meissner state $(\psi = \psi_0)$ with respect to the normal state $(\psi = 0)$. Using this expression the components of the stress tensor $\underline{\underline{\sigma}}_L$ as a function of the distance r from the vortex center can be calculated. The interaction force between a flux line and a dislocation is then given by the PEACH-KOEHLER (1950) formula

$$\underline{K} = \underline{t} \times (\underline{\underline{\sigma}}_L \cdot \underline{b}_o) \tag{3.15}$$

where $\underline{b}_o$ is the Burgers vector and $\underline{t}$ is the unit sense vector of the dislocation. For an edge dislocation parallel to the flux line $(\underline{t}||\underline{B})$ KRAMER and BAUER find a maximum force per unit length

$$\frac{K_o}{L_z} = \frac{\varepsilon_{vo}(1+\nu)}{3S_{44}(1-\nu)} \frac{b_o}{4} \tag{3.16}$$

which occurs if the Burgers vector $\underline{b}_o$ is perpendicular to the line connecting the centers of the vortex and the dislocation. If $\underline{b}_o$ is parallel to this line K_o/L_z is zero. In Eq. (3.16) ν is Poisson's ratio and S_{44} is the shear compliance. Inserting $\varepsilon_{vo} \simeq 2.10^{-7}$, $S_{44} = 3.10^{-11}$ m^2 N^{-1}, $\nu = 1/2$, and $b_o = 3.10^{-10}$ m as typical values into Eq. (3.16) yields a maximum force between a vortex and a parallel edge dislocation of about 5.10^{-7} Nm^{-1}.

There is no first order elastic interaction between a vortex and a parallel screw dislocation since the Burgers vector is perpendicular to the stresses around the vortex core and the product $(\underline{\sigma}_L \cdot \underline{b}_o)$ in Eq. (3.15) vanishes. The same result is obtained for screw dislocations perpendicular to the flux lines and for edge dislocations with $\underline{t}$ and $\underline{b}_o$ perpendicular to $\underline{B}$. Edge dislocations perpendicular to the flux lines but with $\underline{b}_o$ parallel to $\underline{B}$ will give a non-zero but very small interaction force.

Finally for a small spherical inclusion of diameter $L \ll \xi$ and a misfit parameter ξ, KRAMER and BAUER obtained

$$K_o \approx \frac{10L^3 \zeta \varepsilon_{vo}}{S_{44} \xi} \tag{3.17}$$

The dielastic interaction can be visualized as follows. The elastic self energy of a lattice defect depends on the elastic constants of the surrounding lattice and will therefore be different for a defect in the "stiff" vortex core and a defect in the "soft" superconducting surrounding. This leads to a repulsive force between the defect and the flux line which was first estimated by WEBB (1963) for the case of screw dislocations. The stress field of a screw dislocation is pure shear. Therefore only the change ΔS_{44} of the shear compliance between the normal and superconducting state is involved. WEBB approximated the vortex core by holding ΔS_{44} constant within a radius ξ and zero outside it. He then obtains for the maximum interaction force per unit length of a flux line and a parallel screw dislocation

$$\frac{K_o}{L_z} = (\frac{b_o}{2\pi S_{44}})^2 \frac{\pi}{\xi} \Delta S_{44} \tag{3.18}$$

Assuming 4.10^{-15} m^2N^{-1} as a typical value for ΔS_{44}, we obtain a force $K_o/L \approx 5.10^{-6}$ Nm^{-1}. This is about an order of magnitude larger than the parelastic interaction force for a parallel edge dislocation (Eq. 3.16).

For a perpendicular screw dislocation the result is

$$K_o = (\frac{b_o}{2\pi \, S_{44}})^2 \; 3\pi \; \Delta \; S_{44} \qquad\qquad (3.19)$$

The dielastic interaction between a flux line and a parallel
edge dislocation was calculated by KRAMER and BAUER (1967). The
result cannot be expressed in closed form. Numerical evaluation
for typical parameters gives an interaction force which is of
the same order of magnitude as that for the parelastic interaction (Eq. 3.16).

The dielastic effect of spherical inclusions was treated
by TOTH and PRATT (1964). The result is

$$K_o = \frac{3\pi \, L^3 \, \zeta^2}{8 S_{44}^2 \, \xi} \, \Delta S_{44} \simeq \frac{3(2\pi)^{3/2} \; \zeta^2 \; \Delta S_{44}(0)}{16 \phi_o^{1/2} \; H_{c2}(0) S_{44}^2} \; H_{c2}^{3/2} \; L^3 \qquad (3.20)$$

where we have used the experimental finding that ΔS_{44} has a
similar temperature dependence as H_{c2} (ALERS and WALDORF 1961).
$\Delta S_{44}(o)$ and $H_{c2}(0)$ are the values of ΔS_{44} and H_{c2} at $T = 0$. The
quadratic dependence on the misfit parameter ζ indicates that
for large ζ-values the dielastic interaction will dominate
whereas particles with small misfit parameters will mainly pin
via the parelastic interaction which depends linearly on ζ
(Eq. 3.17).

3.2 Treatments Based on the Ginzburg-Landau Theory

In the high field region where the distance between flux
lines becomes comparable to the coherence length ξ, a distinc-
tion between core interactions and magnetic interactions is not
very useful. Since the order parameter is small in this region
(see Fig. A4), the Ginzburg-Landau theory will provide a more
suitable description of the mixed state. In this theory (see
Chapter 8) the properties of the superconductor are characterized by the parameters α and β and we will therefore expect
pinning if structural defects introduce spatial fluctuations of
these parameters.

Neglecting elastic energy terms the change in the free
energy of the mixed state caused by small fluctuations $\delta\alpha$ and
$\delta\beta$ is given by (see Eq. (8.20))

$$\delta E = \int \mu_o H_c^2 \left(- \frac{\delta H_{c2}}{H_{c2}} \left|\frac{\psi}{\psi_o}\right|^2 + \frac{1}{2} \frac{\delta \kappa^2}{\kappa^2} \left|\frac{\psi}{\psi_o}\right|^4 \right) dV \tag{3.21}$$

In this expression Eq. (8.21) was used to replace α and β by
the measurable quantitites H_{c2} and κ. CAMPBELL and EVETTS (1972,
p. 335) applied Eq. (3.21) to calculate the pinning force caused
by a small volume V ($L < \xi$) with a different upper critical field
H_{c2}. An example of such a pinning center is a dislocation tangle
in which the reduced electron mean free path leads to a small increase in H_{c2} (see Eq. 8.29). If the periodic flux lattice is
approximated by its first harmonic (see e.g. ST. JAMES et al.
1969, p. 61) one can show that the order parameter varies from
$|\psi/\psi_0|^2 = 0$ for the pinning center on the vortex core to
$|\psi/\psi_0|^2 = 3 <|\psi/\psi_0|^2>/2$ for the pinning center half way between
the vortices. The maximum pinning energy is then given by

$$U_o = \frac{3}{2} \mu_o H_c^2 V \frac{\delta H_{c2}}{H_{c2}} <\left|\frac{\psi}{\psi_o}\right|^2> \tag{3.22}$$

because the $\delta\kappa^2$ term in Eq. (3.21) can be neglected at high
flux densities where $|\psi|^4 << |\psi|^2$. U(x) has a maximum slope of
$\pi U_o/a_o$ at a distance $x = a_o/4$ from the vortex center which leads

to a maximum pinning force of

$$K_o = \frac{3\pi}{4a_o} \mu_o \left(\frac{H_{c2}}{\kappa}\right)^2 V (1 - b) \frac{\delta H_{c2}}{H_{c2}}$$

Here Eqs. (8.23) and (8.26) were used to replace $|\psi/\psi_o|^2$ and
H_c by (1-b) and H_{c2} and κ, respectively. Since for a hexagonal
lattice $a_o = (2 \phi_o/\sqrt{3} B)^{1/2}$ we finally obtain using $V = \pi L^3/6$

$$K_o = \frac{\pi^2}{8} \left(\frac{\sqrt{3}}{2}\right)^{1/2} \frac{\mu_o}{\phi_o^{1/2}} \frac{H_{c2}^{3/2}}{\kappa^2} \frac{\delta H_{c2}^{5/2}}{H_{c2}} L^3 b^{1/2} (1 - b) \tag{3.23}$$

For large perturbations of α and β which may be caused by a void or an insulating inclusion in the material, it would be necessary to solve the Ginzburg-Landau equation with the corresponding boundary conditions at the pinning center. No calculations of this kind have been reported up to now. However, if the void is very small ($L \ll \xi$), Eq. (3.23) might still be a reasonable approximation since a small void at the vortex core or half way between vortices will not change the structure of the mixed state very much. Indeed, setting $\delta H_{c2}/H_{c2} = 1$ yields an expression which seems to connect the low field value Eq. (3.2) and the point $K_0 = 0$ at H_{c2} in a reasonable way.

If the variations of α and β are caused by stress fields (elastic interaction) the calculation of the pinning energy is much more difficult. Perturbational calculations of the generalized Ginzburg-Landau equations which are derived by minimizing the free energy of the mixed state including elastic terms were first carried out by LABUSCH (1968) and KAMMERER (1969). MIYAHARA et al. (1969) confirmed Labusch's result and give explicit expressions for some types of defects (point defects, screw dislocations and edge dislocations). An even more general approach was attempted by SEEGER and KRONMÜLLER (1968) and KRONMÜLLER and SEEGER (1969) who included strain gradient terms and non-linear effects. The resulting micromagnetic equations of a deformable superconductor are rather complex and contain a large number of parameters characterizing both the structure of the mixed state and the elastic defect. Although LABUSCH (1968) has given relations connecting some of these parameters with experimental accessible quantitites, reliable measurements are still scarce. Therefore, a comparison of the above theories with experiments is hardly possible at present. For this reason these calculations will not be discussed here and the theory-minded reader is referred to the original papers or to the review of CAMPBELL and EVETTS (1972). It should only be mentioned that, because of a dominance of the $|\psi|^2$ term, the elastic pinning interactions obtained from Ginzburg-Landau calculations decrease roughly as (1-b) when the flux density approaches the upper critical field (see Eq. 8.26). This is a similar field dependence as for the δH_{c2}-interaction in Eq. (3.23), which indicates that the field dependence of interaction forces caused by different pinning mechanisms will look rather alike: independent of b at low flux densities (b → 0), and decreasing linearly to zero at B_{c2} (b → 1) (see Fig. 26). This shows that a measurement of pinning forces at only one temperature is usually not sufficient to draw any conclusions about the pinning mechanism (s) involved. However, the temperature dependence of K_o can be quite different for different mechanisms (e.g. compare Eqs. (3.2) and (3.20)) and can thus be very valuable for the identification of pinning interactions (see also Chapter 5).

4. Relations between Macroscopic and Local Forces

After the discovery of high field hard superconductors in 1961, a rapidly growing number of papers, describing measurements of critical current densities J_c in various materials was published. The results clearly demonstrated the importance of lattice defects in the current carrying capacity of hard superconductors. A drawback of this early work, however, was that the metallurgical structure of most samples was unknown or only vaguely known. This state of affairs prevented an analysis of the measurements and led to the unfortunate practice of "interpreting" the experimental results by so-called pinning models (which were just different mathematical expressions for fitting the field dependence of J_c).

Probably the first attempt to relate flux pinning results quantitatively to individual flux line-defect interactions was that of NEMBACH (1966). Using theoretical expressions for the elastic interaction between a dislocation and a vortex line (WEBB 1963, see Chapter 3) he found that the measured macroscopic pinning force density P_v was more than an order of magnitude smaller than the algebraic sum of all maximum individual forces per unit volume. Similar results were obtained by ALDEN and LIVINGSTON (1966) who studied the pinning effects of ferromagnetic particles.

LABUSCH (1966b) suggested that this discrepancy is a result of the strong coupling of the flux lines to each other which prevents them from adjusting to the array of pinning centers in an optimum way. Independently, a similar conclusion was reached by YAMAFUJI and IRIE (1967) who showed that the energy dissipation in hard superconductors can only be explained if

the influence of the flux lattice rigidity on the pinning process is taken into account.

Theoretical calculations based on these ideas are in reasonable agreement with recent experiments on samples with known
and well defined defect structures. It therefore seems that the
present theories which will be discussed in the following, are
able to provide a suitable basis for an understanding of flux
pinning in type II superconductors. It should not be overlooked,
however, that there are still open questions (as e.g. the problem of the pinning threshold, see Sect. 4.2.3).

4.1 The Summation of Pinning Forces in a Simple Model

In the critical state the driving force density P_D is balanced by the pinning force density P_V.

$$\underline{P}_D + \underline{P}_V = O \tag{2.20}$$

The macroscopic force P_V represents the combined action of a
large number of individual interaction forces K between vortices and pinning centers. If $\rho(K)dK$ is the number of interactions
per unit volume with forces between K and K + dK, P_V can be
written as

$$P_V = \int_{-K_o}^{+K_o} K \, \rho(K) \, dK \tag{4.1}$$

K_o is the maximum force a pinning center of a given type can
exert on a flux line. In general $\rho(K)$ depends on the strength
and distribution of the pinning centers and on the distortions
they are able to produce in the flux line lattice. Therefore,
a straightforward evaluation of the integral (4.1) is possible
only for two limiting cases:

(a) If the vortex lattice is completely rigid and if there
is a sufficiently large number of statistically distributed pinning centers, the number of interactions with force K will be

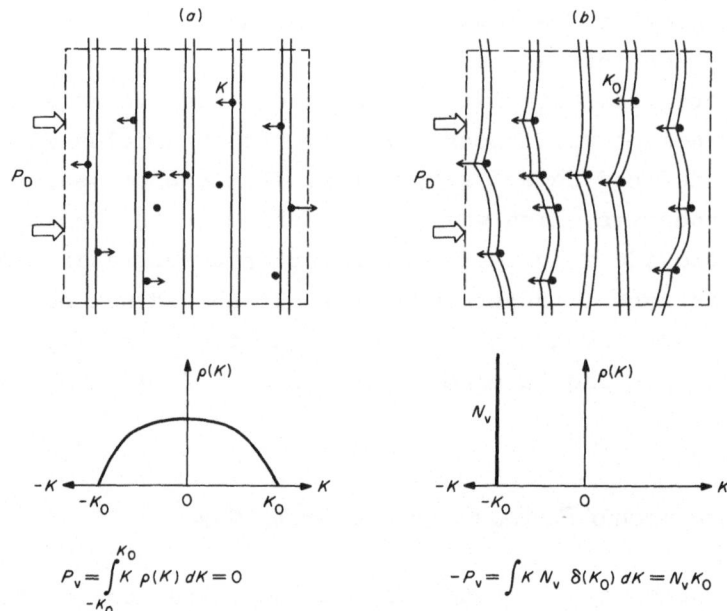

$$P_v = \int_{-K_0}^{K_0} K \, \rho(K) \, dK = 0 \qquad\qquad -P_v = \int K \, N_v \, \delta(K_0) \, dK = N_v K_0$$

Fig.22 Unit volume containing a statistical array of pinning centers interacting with (a) a completely rigid flux lattice and (b) an almost completely soft lattice. The density of the pinning sites is N_v and their maximum interaction force is K_0.

equal to that with force $- K$, i.e. $\rho(K)$ is an even function (Fig. 22a). The integral (4.1) then vanishes and we have

$$P_v \to 0 \quad \text{for} \quad C_{ij} \to \infty \tag{4.2}$$

(b) On the other hand in an almost completely soft vortex lattice the flux lines can adjust to the pinning array in an optimum way. Thus, if there are N_v pinning points per unit volume which can exert a maximum force K_0 on each flux line, we have (Fig. 22b)

$$P_v \to -N_v K_0 \quad \text{for} \quad C_{ij} \to 0 \tag{4.3}$$

Here we have assumed that not more than one flux line occupies any one pinning site. It can be expected that the direct summation of individual interaction forces leading to Eq. (4.3) may also be valid for a real flux lattice if the pinning forces are

sufficiently strong to completely disrupt the lattice. DEW-
HUGHES (1974a) suggests that this is the case in most high-K
hard superconductors and flux lattice rigidity effects may
therefore be ignored in these materials.

   If, on the other hand, the interaction energy between pinning centers and vortices is small compared to the self-interaction between vortices, the evaluation of the integral (4.1)
for the three-dimensional case is a difficult task. In order to
get a feeling about the summation process it is therefore more
appropriate to analyze first a simple two-dimensional model.
Such a calculation still reflects most of the essential features of the general case and has the advantage of being more
lucid.

   In this model which is similar to the one introduced by
LOWELL (1972) a pinning center is characterized by a potential well $U(x)$ as shown in Fig. 23a. The corresponding force
$K = - dU/dx$ on a flux line within this potential well will lead
to a distortion u which causes an elastic counter force $- C'u$.
The equilibrium distortion $u_0(x_0)$ for a flux line at position
x_0 in the undistorted lattice (i.e. far above or below the pinning center, see Fig. 23c) is given by

$$C' u_0(x_0) = \frac{dU}{dx}\bigg|_{x_0+u_0} = \frac{dU}{dx}\bigg|_{x_0} + \frac{d^2U}{dx^2}\bigg|_{x_0} u_0$$

$$u_0(x_0) = \frac{dU/dx\big|_{x_0}}{C'-d^2U/dx^2\big|_{x_0}} \tag{4.4}$$

C' is the appropriate elastic constant and d^2U/dx is the rate
of change of the force K with distance. Equation (4.4) remains
valid only as long as the denominator stays finite, i.e.
$C' > d^2U/dx^2$ (dashed line A in Fig. 23b). In the case of a stable lattice it is immediately seen from Eq. (4.4) that

$$- u_0(x_0) = u_0(-x_0) \tag{4.5}$$

leading to the distortions shown schematically in Fig. 23c.
This pattern is symmetrical on the two sides of the pinning

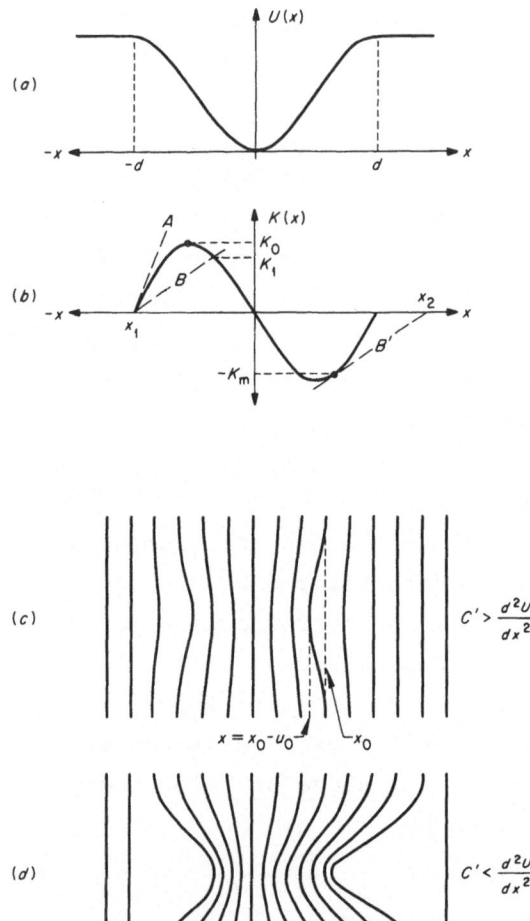

Fig.23 (a) Free energy U of a flux line passing through a typical pinning center as a function of the flux line position x. (b) Interaction force $K = - dU/dx$ as a function of line position x. The slopes of the dashed lines characterize flux lattices with different elastic constants C'. Line A shows a rather stiff lattice with $C' > d^2U/dx^2$ which is stable throughout the whole pinning range, 2d. A soft lattice (lines B and B') is elastically unstable in regions where $C' < d^2U/dx^2$. (c) and (d) are a schematic indication of the shape of a flux line passing slowly through the pinning center in + x direction. Case (c) corresponds to the stiff lattice (line A in b) where the distortion u_o is introduced continuously and is symmetrical on the two sides of the pinning center. The softer lattice in (d) is unstable in certain regions causing discontinuous and asymmetric distortions. It should be emphasized that the distortions u_o in this figure are strongly exaggerated. In reality the maximum angle between a distorted line element and the external field direction will be in the order of 1° (see Section 7.5 and Fig. 62).

site and the distortions are introduced continuously as the flux line is moved across the potential well. Since symmetrical distortions are equivalent to symmetrical forces, the interaction density $\rho(K)$ will be an even function and no net force will be exerted on the flux lattice:

$$P_V = O \quad \text{for} \quad d^2U/dx^2 < C' \qquad (4.6)$$

On the other hand, if the rate of change of K becomes equal to or larger than C', an elastic instability will develop. The position x where the instability occurs depends on the direction of the driving force which is schematically shown in Fig. 23d. As soon as a flux line moving to the right (driving force in +x direction) reaches the position $x_1 = - d$ the force due to

the pinning center increases more rapidly than the maximum
elastic counter force (dashed line B in Fig. 23b). Therefore
the flux line will not come to a halt until it reaches a region
where C' > d^2U/dx^2 and where $K = K_1$ has appreciably dropped below its maximum value K_o. For positions $x > 0$ however, an instability does not occur until $x_o = x_2 > d$ is reached (dashed
line B' in Fig. 23b). It is precisely this asymmetry which
gives rise to a net pinning force because there are now more
interactions with forces - K (right half of the pinning center)
than with forces + K (left hand). The interaction density $\rho(K)$
is now an asymmetrical function as shown in Fig. 24. The exact
form of $\rho(K)$ will depend on the details of the potential well
of the pinning center but for our present qualitative discussion
may be approximated by the dashed lines[+] in Fig. 24, i.e.

$$\rho(K) = \rho(0) \qquad -K_o < K < K_1$$

$$\rho(K) = 0 \qquad K > K_1; \quad K < K_o \tag{4.7}$$

$\rho(0)dK$ can be calculated in the following way: if the flux
lattice is displaced by a small distance dx_o, each flux line
will contact $N_v\bar{L}_y dx_o$ new pinning centers per unit length. N_v
is the density of pinning centers and $\bar{L}_y$ is their effective
length in the direction perpendicular to the flux lines and the

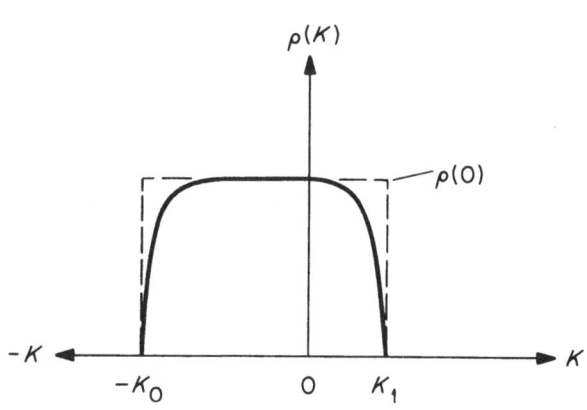

Fig.24 Number of interac-
tions $\rho(K)$ per unit volume
with forces between K and
K + dK for the case (d) in
Fig.23. Since no vortices
are found in the unstable
region in the left side of
the pinning site there are
no interactions with forces
larger than $K_1 < K_o$. However, on the right side the
pinning center is able to
exert forces up to - K_o on
the flux line. The dashed
line shows the approxima-
tion for $\rho(K)$ used in the
derivation of Eq.(4.9).

[+]Different functions $\rho(K)$ will change the final result (4.10) only by a
numerical factor but will leave its general form unaltered.

driving force (for very small pinning sites of diameter $L_y << d$, $\bar{L}_y$ is approximately equal to $2d$; for large sites with $L_y > d$, $\bar{L}_y \simeq L_y + 2\,d$). At a flux density B there are B/ϕ_0 vortices per unit volume and we therefore obtain

$$\rho\,(0)\,dK = \frac{B}{\phi_0}\; N_v\,\bar{L}_y\; dx_0 \qquad\qquad (4.8)$$

The displacement dx_0 is equivalent to a force change $dK = C'du \simeq C'dx_0$ and with Eq. (4.1) we finally obtain for the macroscopic pinning force density

$$P_v = \frac{B}{\phi_0}\frac{N_v\,\bar{L}_y}{C'} \int_{-K_0}^{K_1} K\;dK = -\frac{B}{\phi_0}\frac{N_v\,\bar{L}_y}{C'}\,(K_0^2 - K_1^2) \qquad (4.9)$$

The minus sign indicates that P_v acts in the $-x$ direction thus opposing the driving force P_D which we assumed to be in $+x$ direction. For the case of energy well curvatures d^2U/dx^2 much larger than the stiffness C' of the vortex lattice, K_1 will be much smaller than K_0 (see Fig. 23b) and Eq. (4.9) reduces to

$$P_v \simeq -\frac{B}{\phi_0}\frac{N_v\,\bar{L}_y}{C'}\,K_0^2 \qquad\qquad \text{for } d^2U/dx^2 >> C' \qquad (4.10)$$

The expressions (4.6) and (4.10) for the bulk pinning force density P_v show several important features. First, for a given flux lattice, we expect substantial pinning only from pinning centers with sufficiently sharp potential wells which are able to create elastic instabilities in the vortex lattice. Second, if this criterion is fulfilled, P_v appears in both the cases when the flux line-pin interaction is either attractive or repulsive, because $P_v \propto K_0^2$. Third, the pinning force density increases with decreasing rigidity of the flux lattice.

Historically, a relation of the type (4.10) was first derived in 1967 by YAMAFUJI and IRIE using a rather different approach. They considered the energy dissipated when flux lines are released from their pinning sites and equate this to the work done by the pinning force density (for a short discussion on energy dissipation of moving vortices see Chapter 6). It is evident that this approach also requires elastic instabilities in order to obtain a non-zero pinning force, i.e. if only continuously introduced distortions occur there will be no dissipation and therefore no bulk pinning force. Calculations of this kind are usually referred to as the dynamic approach as distinguished from the statistical approach used above where one counts the number of interactions on the two sides of the pinning well.

4.2 Results and Validity Limits of Present Pinning Theories

4.2.1 Point forces

The first thorough calculation of the bulk pinning force P_V caused by a statistical array of pinning centers is due to LABUSCH (1969a). A detailed presentation of Labusch's theory would be beyond the scope of this article. Therefore only the conditions for its validity and the final result for a specific pinning potential shall be given here. During the derivation of his statistical theory Labusch made the following assumptions:

(a) The interaction between flux line elements is described by linear response theory. This means that the distortions in the flux lattice have to be small or, in other words, the interaction energy U must be weak compared to the self-interaction between vortices. This is the case of the lattice approximation which was discussed in Sect. 2.2.1 (see Eq. 2.58).

(b) The pinning centers exert point forces on the flux lines and the displacements in the lattice are given by Eq.(2.48).

(c) The pinning array is dilute, i.e. the total interaction energy can be expressed as the sum over two-particle potentials (criteria for dilute arrays are given by Eqs. (4.23) and (4.24)).

(d) The pinning array is random, i.e. the pinning centers do not act cooperatively to increase the displacement above the value given by Eq. (2.48).

For an explicit evaluation of his general result, Labusch chose a pinning potential U of the following form:

$$U = U_o \left(1 - \frac{2}{3} \frac{x^2}{d^2} + \frac{1}{9} \frac{x^4}{d^4}\right) \qquad \text{for } |x/d| < \sqrt{3}$$

$$U = 0 \qquad \text{for } |x/d| > \sqrt{3}$$

(4.11)

and he found that a density N_v of defects with this potential cause a pinning force density P_v of

$$P_v \simeq \left(\frac{B}{\phi_o}\right) N_v \bar{L}_y \frac{G'(0)}{2} K_o^2 \qquad \text{for } \frac{K_o G'(0)}{d} > \frac{8}{3}$$

$$P_v \simeq 0 \qquad \text{for } \frac{K_o G'(0)}{d} < \frac{2}{3}$$

(4.12)

For $2/3 < K_o G'(0)/d < 8/3$, P_v is given by a more complicated function which is graphically shown in Fig. 25. In Eq. (4.12) $\bar{L}_y$ is again the effective length of the pinning center in the direction perpendicular to the field and to the driving force, and U_o has been replaced by the maximum interaction force $K_o = \pm 8 U_o/9d$. $G'(0)$ is a function describing the distortion

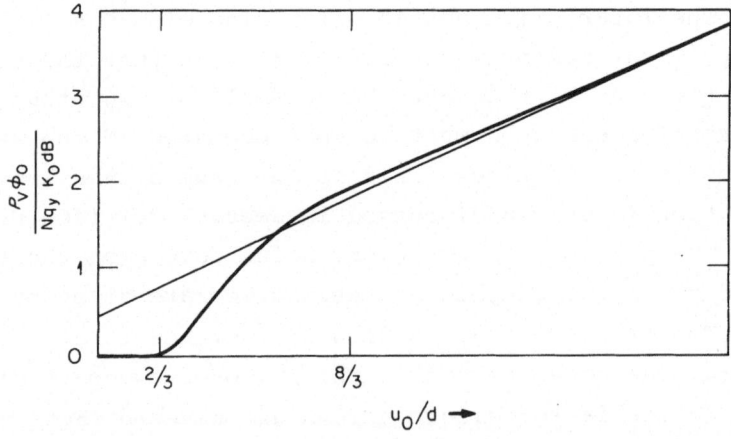

Fig.25 Ratio between pinning force density per defect and interaction strength as a function of the relative distortion u_o/d. Although the curve was calculated for the specific potential of Eq.(4.11) it may be taken as representative for an arbitrary potential characterized by a maximum interaction force K_o and a width d (from LABUSCH 1969a).

u_o of a flux line in a lattice due to a point force and is given in Eq. (2.48) in terms of the elastic constants of the flux lattice. Inserting Eq. (2.48) into Eq. (4.12) we finally obtain for the pinning force caused by a dilute array of point forces:

$$P_v = N_v \bar{L}_y \left(\frac{B}{\phi_o}\right)^{3/2} \frac{1}{8\sqrt{\pi}} \left(\frac{1}{(C_{44} C_{66})^{1/2}} + \frac{1}{(C_{44} C_{11})^{1/2}}\right) K_o^2 \qquad (4.13)$$

for $u_o/d > 2/3$

Apart from a factor 2, this expression is identical with Eq. (4.10) if we set $C' = 1/G'(0)$ and the condition for the pinning threshold is also quite similar since

$$\frac{1}{C'} \frac{d^2 U}{dx^2} \simeq \frac{1}{C'} \frac{K_o}{d} = \frac{G'(0) K_o}{d} > \frac{8}{3} \qquad (4.14)$$

Using the same assumptions (a) - (d) as above and requiring that the condition

$$\frac{dK}{dx} > \frac{1}{G'(0)} \simeq 4 \left(\frac{\pi \phi_o C_{44} C_{66}}{B}\right)^{1/2} \qquad (4.15)$$

(see Eq. 2.49) is fulfilled over most of the range of the pinning center, CAMPBELL and EVETTS (1972, p. 312) have derived the pinning force density with the dynamic approach. The result is identical with Eq. (4.13) which indicates that in general we may expect the dynamic pinning force to be the same as a summation of static pinning forces.

4.2.2 Line and area forces
    The first calculation of the pinning force density caused by an array of line forces parallel to the vortices was carried out by GOOD and KRAMER (1970) using the dynamic approach. They obtain

$$P_v = N_v L_z \left(\frac{B}{\phi_o}\right)^{1/2} \frac{5}{16\pi} \frac{1}{C_{66}} \left(\frac{K_o}{L_z}\right)^2 \ln \left(\frac{B}{4N_v L_z \phi_o}\right)^{1/2} \qquad (4.16)$$

where L_z is the length of the pinning lines and N_v is their volume density. In the case of line forces the distortion is given by Eq. (2.45). Therefore only the shear modulus C_{66} appears in Eq. (4.16) and in the pinning threshold which is given by

$$\frac{1}{L_z} \frac{dK}{dx} > \frac{16\pi\, C_{66}}{5\, \ln(B/4\, N_v\, L_z\, \phi_0)^{1/2}} \qquad (4.17)$$

A calculation of P_v for an array of line forces which are not parallel to the flux lines and which would be of more practical interest is difficult and has not yet been attempted.

Pinning by plane boundaries parallel to the vortices was discussed qualitatively by CAMPBELL and EVETTS (1972). They point out that in general both the statistical and the dynamic approach are difficult to apply to this problem, mainly because the distortion caused by an area force depends on the gauge length L_x (see Eq. 2.44). A simple expression for P_v can only be given for the case where the plane spacing is sufficient for the planes to act independently. This means that the distortions between planes must be large enough to allow two neighboring planes to exert their maximum force K_{max} on the flux lines. If there are N_v planes per unit volume the bulk force density is then simply given by $P_v = N_v K_{max}$. For comparison with experimental results it is convenient to replace the total force K_{max} per plane by the force $K_0/L_z = K_{max} a_0/A$ per unit length of a single flux line. With this we obtain

$$P_v = N_v\, A\, \left(\frac{B}{\phi_0}\right)^{1/2} \frac{K_0}{L_z} \qquad (4.18)$$

where A is the area of the planes perpendicular to the driving force and L_z is their dimension parallel to the vortices.

4.2.3 Is the theoretical pinning threshold too high?

The preceding discussion showed that all present pinning theories give a non-zero bulk pinning force only if the interaction forces are larger than a certain threshold value[+]. It is therefore rather important to see whether the interaction for-

64

ces from the different pinning mechanisms listed in Chapter 3 are able to meet the threshold criteria. For such a comparison it is useful to rewrite Eqs. (4.15) and (4.17) in terms of the experimental parameters $b = B/B_{c2}$, B_{c2}, and κ. Using Eqs. (2.42) and (2.43) we find for fields not too close to H_{c1}

$$\frac{dK}{dx} > \frac{2.5\phi_0^{1/2} \; B_{c2}^{3/2}}{\mu_0 \quad \kappa} \; b^{1/2} \; (1 - b) \tag{4.20}$$

for point forces and

$$\frac{1}{L_z} \frac{dK}{dx} > \frac{1.3 \; B_{c2}^2}{\mu_0 \quad \kappa^2} \; \frac{1}{\ell n (B/4N_v L_z \phi_0)^{1/2}} \; (1 - b)^2 \tag{4.21}$$

for line forces. Figure 26 shows that the field dependence for point forces is similar to that of most interaction forces (see Chapter 3). Thus, weak point pins which are unable to meet condition (4.20) will be ineffective over the whole field range

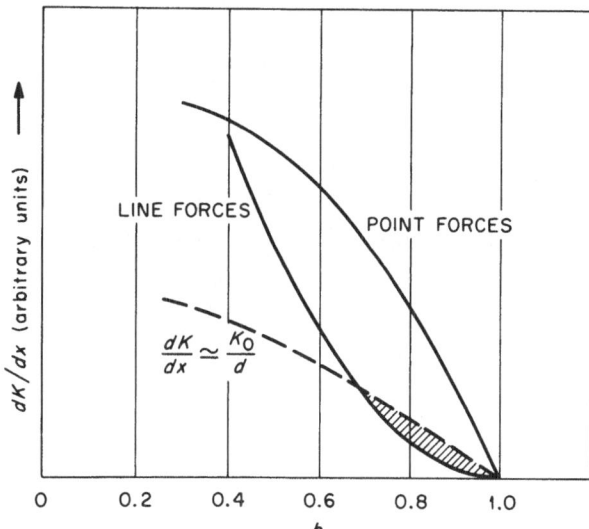

Fig.26 Field dependence of the pinning threshold for point and line forces, respectively. The scale for dK/dx is arbitrary. The dashed line represents the field dependence of dK/dx for a typical (weak) pinning center.

Footnote, p. 64:

[+]This applies for the present "first order" theories. n[th]-order calculations where the density function $\rho(K)$ depends on the positions of a vortex relative to n neighboring defects can make the actual bulk force different from zero although the first order calculation has a zero result (LABUSCH 1969a).

(except very close to H_{c1}). The pinning threshold for lines, however, goes nearly quadratically to zero at H_{c2}, which is much faster than most interaction forces, and there will be a high field region (shaded area in Fig. 26) where even weak line forces may become effective pinning centers. This can be one of the reasons for the peak effect (see next section) observed in some materials.

If typical parameters for various pinning centers are inserted into the theoretical expressions for K_o in Chapter 3, one usually finds dK/dx values which are by far too small to meet the threshold criteria (4.20) and (4.21), respectively. To illustrate this in an example we calculate K_o for a large void in our "standard" superconducting matrix (see Chapter 3) with B_{c2} = 12 T and κ = 50 at 4.2 K. Assuming that this very strong pinning center has a linear dimension parallel to the vortices of L_z = 1000 Å we obtain from Eq. (3.4) $K_o \simeq 2 \cdot 10^{-11}$ N. The lower limit for the interaction distance d is the coherence length $\xi \simeq 50$ Å, i.e. the maximum attainable value for dK/dx $\simeq K_o/\xi$ will be about $4 \cdot 10^{-3}$ Nm^{-1}. This is about an order of magnitude smaller than the threshold value of $3 \cdot 10^{-2}$ Nm^{-1} calculated from Eq. (4.20) for b = 0.5. Experimentally, however, it is found that even much weaker pinning centers will pin as effectively as if there was no threshold criterion.

An explanation for this discrepancy was attempted by FIETZ and WEBB (1969). They consider the statistics of clusters of m pinning centers which combine to produce an interaction force mK_o per cluster. Flux creep experiments (BEASLEY et al. 1969) from which the total pinning energy and the volume of a cluster can be determined, support this view. In a random array such clusters occur with a frequency $N_v \bar{L}_y/m^2$ per unit volume. Only groups with a critical size n_c = m_c^2 given by the condition that m_cdK/dx will meet the threshold criterion, are expected to contribute to P_v. Therefore, instead of Eq. (4.12) the bulk pinning force will be given by

$$P_v = (\frac{B}{\phi_o}) \frac{N_v \bar{L}_y}{m_c^2} \frac{G'(0)}{2} (m_c K_o)^2 \qquad (4.22)$$

which is identical to Eq. (4.12) because the group size m_c^2 cancels out. Equation (4.12) therefore should yield the correct pinning force density even for the case where the individual pinning centers do not fulfill the threshold criterion. For weak pinning centers the necessary cluster size is rather large, e.g. for dislocations in Pb-Tl alloys, BEASLEY et al. (1969) find $n_c \approx 10^2 - 10^4$. Even at high concentrations the volume V occupied by so many dislocations is rather large and the assumption of point forces is certainly no longer valid.

Another possibility to explain the low observed threshold values was pointed out by BIBBY (1970). If the interaction force is exerted on the vortex core, the core might be displaced from the geometrical center of the flux line. This would lead to stronger distortions as calculated from the elastic constants where the cores were regarded as fixed in the middle of each line. Another reason for a softening of the flux lattice could be the presence of large concentrations of flux lattice defects (see Sect. 2.2.2). According to CAMPBELL and EVETTS (1972, p.394) this effect is most likely at high flux densities and might be one of the reasons for a peak effect.

Although these ideas are undoubtedly important for many cases, no really satisfactory general way of explaining the low threshold has yet been found. It is interesting to note that the same problem arises in the description of mechanical hardening of metals by dislocation pinning (LABUSCH and HAASEN 1973). Indeed, many other similarities in the description of mechanical, magnetic, and superconducting "hardness" have been emphasized by HAASEN (1970) and it is hoped that future progress in one of these fields will be beneficial to the others too.

4.3 Concentrated Arrays, Peak Effects and Other Phenomena

In this section I shall discuss some phenomena which cannot be described by the theoretical treatments of the preceding section, as e.g. nondilute or nonrandom pinning center distributions. However, before starting to investigate concentrated arrays a more quantitative criterion for "dilute" should be given. In order to meet condition (c) in Section 4.2.1, the strains in the vortex lattice introduced by the individual pinning cen-

ters should not overlap considerably. If we require that the
distortions of the lattice caused by a neighboring center are
less than one-tenth of the distortions at a given pinning center, a criterion for "dilute" would be that the distance ℓ between point forces is larger than 10 vortex spacings a_o (see
Eq. 2.47). Thus, an array of large pinning centers is dilute if

$$N_V < 10^{-3} \left(\frac{B}{\phi_o}\right)^{3/2} \tag{4.23}$$

where $N_V = 1/\ell^3$ is the concentration of the pinning centers.

For small centers ($L_y \simeq 2d << a_o$) however, condition (4.23)
is too stringent since, as was pointed out by CAMPBELL and
EVETTS (1972, p. 313), the probability that two neighboring
pins are interacting simultaneously with the vortex lattice is
only about $(2d/a_o)^2$. This is again a consequence of the rigidity of the flux lattice which prevents the flux lines from adjusting to the pinning array completely. For small centers we
therefore have

$$\left(\frac{2d}{a_o}\right)^2 N_V < \frac{1}{(10a_o)^3} \qquad \text{or}$$

$$N_V < \frac{10^{-3}}{4d^2} \left(\frac{B}{\phi_o}\right)^{1/2} \tag{4.24}$$

If N_V approaches or exceeds the values given by the above conditions, Eq. (4.13) will gradually become inaccurate since neighboring pinning centers will restrict the free volume for displacement and the distortion u_o at a given center will be smaller
than calculated by Eq. (2.48). How much the distortions will be
diminished depends on the distance of neighboring centers, i.e.
the individual forces K for a concentrated array will depend on
the pinning center concentration N_V and we expect that the linear relation between the bulk pinning force P_V and N_V (as found
for the dilute case) will no longer hold.

Concentrated pinning center arrays are observed rather frequently and are probably very important in the commercial materials Nb-Ti and Nb_3Sn but a quantitative calculation of the pinning force density P_V becomes then very complex and has not yet

been attempted for the three-dimensional case. However, some insight into the relationship between P_v, N_v and K_o can be obtained from a one-dimensional model considered by CAMPBELL and EVETTS (1972, p. 317). Assuming very narrow pinning centers each capable of exerting a force up to K_o/L_z per unit length and spaced a mean distance ℓ apart they found

$$P_v = \frac{1}{\ell^{1/2}} \left(\frac{B}{\phi_o}\right) \frac{1}{(16\pi)^{1/2}} \frac{1}{C_e^{1/2}} \left(\frac{K_o}{L_z}\right)^{3/2} \tag{4.25}$$

where $C_e = C_{11}$ is the modulus of the lattice. A physical situation for which Eq. (4.25) should be applicable is a randomly spaced dense array of sharp phase boundaries arranged parallel to the flux lines and perpendicular to the driving force. Very recently APPLEYARD et al. (1974) and EVETTS (1974) carried out numerical calculations of pinning by one and two-dimensional arrays of identical pinning centers with densities ranging from 0.1 to about 30 per flux line. They employ three different methods ("complete solution, push-up method and landslide approach") to compute the bulk pinning force and good agreement between the methods is obtained. The calculations for the one-dimensional case confirm Eq. (4.25) for large samples. For small samples (L<35 a_o) however, P_v increases anomalously. This effect is expected to occur also in two- and three-dimensions and could be important in fine multifilament Nb-Ti wires (see Section 5.2). Indeed measurements by MAILFERT and PECH (1972) do show a low field anomaly in 2 μ thick wires which may be evidence for this "size" effect.

Preliminary calculations applied to two-dimensional systems yield a pinning force density which is proporational to $N_A^{3/4} \cdot K_o^2$ in contrast to $P_v \propto N_v \cdot K_o^2$ found for the dilute case Eq. (4.12). (N_A is the number of pinning centers per unit area.) Refined computations are under way (EVETTS 1974) and it is hoped that their results will give some insight into the problems of flux pinning by both random and regular dense arrays of pinning centers (as e.g. pinning threshold, peak effects, etc.)

Although one must keep in mind that Eq. (4.25) was derived for a one-dimensional model and the result will probably be some-

what different for the three-dimensional case, it is instruc-
tive to follow the concentration dependence of P_V given by
Eqs. (4.18) and (4.25), respectively. A numerical example of
the normalized pinning force density as a function of $1/\ell$ for
our "standard" matrix (see Chapter 3) at B = 0.1 T is given in
Fig. 27. In the dilute limit ($\ell \gg a_0$) where every pinning
plane acts separately with its maximum interaction force K_0/L_z,
P_V is proportional to $1/\ell$ and Eq. (4.18) will apply (this relation can easily be adopted to one-dimensional problems by
assuming that the pinning planes with area A are infinitely
large, i.e. $N_V \cdot A = 1/\ell$). If their concentration increases, a
growing portion of pinning centers will become ineffective because the distortions in the lattice between the centers now
become too small to place each flux line in its optimum posi-

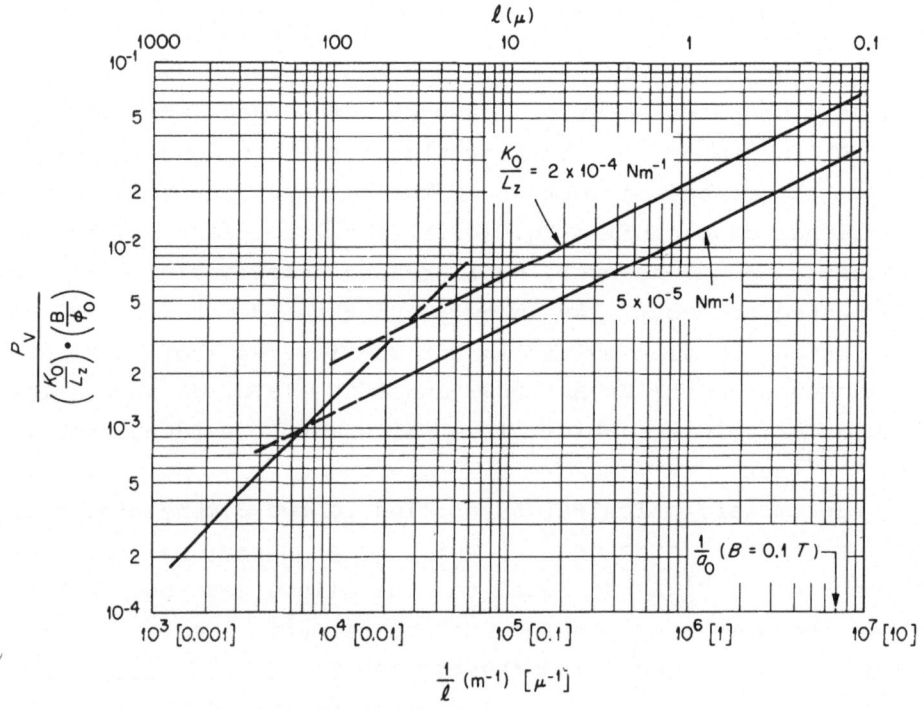

Fig.27 Dependence of the normalized pinning force density on the distance
ℓ between pinning planes in a one-dimensional model for B = 0.1 T. The
curves were calculated from Eqs.(4.18) and (4.25) using our "standard"
matrix parameters. The larger K_0/L_z value corresponds to the interaction
force caused by a sharp void surface as given by Eq.(3.4).

tion. Therefore P_v will increase less than linearly with $1/\ell$ and finally P_v will (at least for the one-dimensional case) be proportional to $1/\ell^{1/2}$ as given by Eq. (4.25). It is clear that the changeover from $1/\ell$ to $1/\ell^{1/2}$ behavior will occur the later the larger the maximum distortion becomes, i.e. the larger K_o/L_z becomes (see Fig. 27). The reason that the one-dimensional system behaves "concentrated" up to $\ell \stackrel{\sim}{<} 200\ a_o$ lies in the use of the elastic constant C_{11}. In the three-dimensional case the distortions are governed by the constants $(C_{44}C_{66})^{1/2}$ and C_{66}, respectively, which are much smaller than C_{11} (see Section 2.2.1) and the criteria for "dilute" will be given by Eqs. (4.23) and (4.24) rather than by the changeover points of the one-dimensional example in Fig. 27.

So far, all the above considerations were based on the assumption of randomly distributed pinning centers. For periodic arrays the situation is quite different because it is now possible that a matching between vortex distances a_o and pinning center spacings ℓ occurs. At flux densities where this is the case, all pinning centers can act cooperatively and produce a bulk pinning force which is much larger than in the general case where the lattice must be distorted in order to give a non-zero pinning force. Such peaks at flux densities which are compatible with the matching condition $\ell = i\ a_o$ (i = 1,2,3 ...) have indeed been observed experimentally (PETERMANN 1970, HILL-MANN and HAUCK 1972). An example is shown in Fig. 28 for an ordered array of α-precipitates in a commercial Nb-Ti alloy: if the precipitate spacing ℓ is equal to multiples of the flux line distance (points A,B,C), a maximum in the $P_v(B)$-curve appears. Peaks which are caused by the matching effect are easily identifiable because the flux densities where they appear are independent of the temperature.

However, maxima in P_v (mostly just below H_{c2}) are also observed in some materials with random pinning arrays and were the subject of numerous experimental investigations. Today, it is clear that there are several mechanisms beside matching that can lead to a "peak effect". The most probable ones are:

(a) A peak in the $P_v(B)$-curve will arise if the basic interaction force K_o exhibits a minimum at some intermediate

71

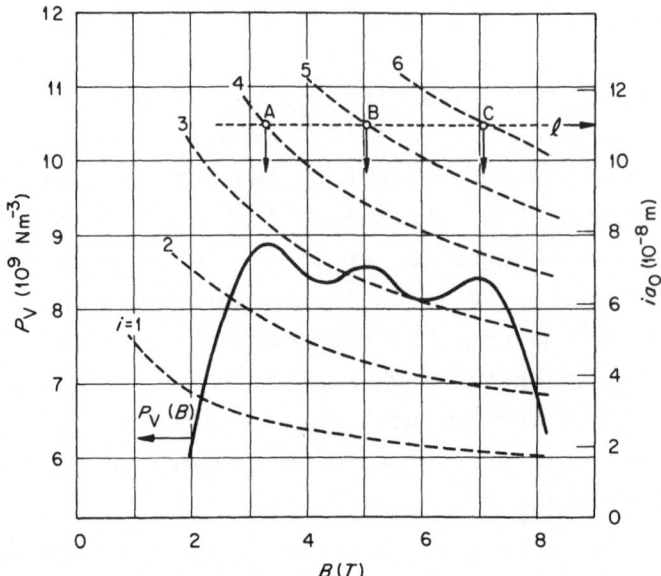

Fig.28 Pinning force density P_V as a function of flux density B in $Nb_{50}Ti_{50}$ containing an ordered array of alpha-precipitates. If the precipitate spacing ℓ matches multiples of the flux line distance a_o (points A,B,C), maxima in P_V appear. The dashed lines give the multiples of a_o as a function of B (after HILLMANN and HAUCK 1972).

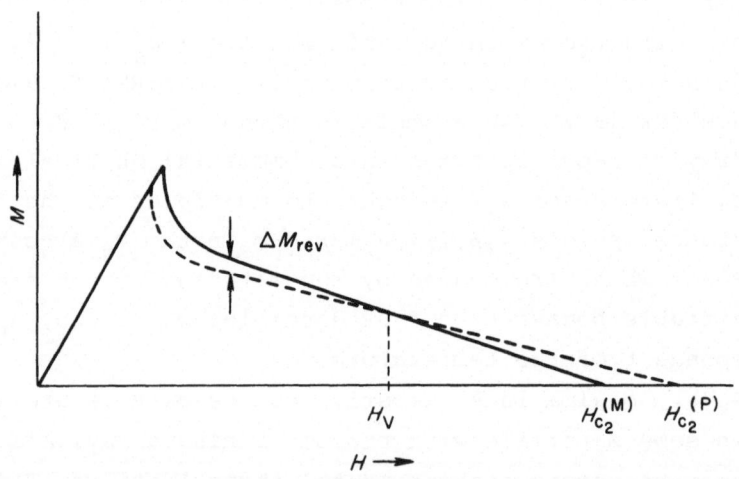

Fig.29 Reversible magnetization curves of matrix (M) and second phase (P). The magnetic pinning interaction which is proportional to ΔM_{rev} vanishes at the field H_V leading to a broad peak in the bulk pinning force somewhere between H_V and $H_{c2}^{(M)}$.

field. This could, for example, be the case for the magnetic
pinning interaction at boundaries between two superconducting
phases with different magnetization curves (Fig. 29). Since
the interaction force is proportional to ΔM_{rev} (see Section
3.1.2) the pinning will go to zero at $H = H_v$ and increase again
for $H > H_v$ until it finally vanishes at H_{c2}. A peak effect of
this kind (which should better be called "valley effect") was
identified by EVETTS and WADE (1970) in Pb-In alloys and by
COOTE (1970) in deformed Nb-Ta alloys. If the temperature is
varied peaks due to a minimum in the interaction force occur at
the same reduced field b rather than at the same flux density B
as in the case of matching. This again stresses the importance
of measurements at different temperatures for the identification
of pinning processes.

(b) In the case of line forces the pinning threshold (4.21)
falls to zero quadratically as B_{c2} is approached whereas most
pinning interactions tend to go to zero linearly. Thus, weak
line forces which are unable to fulfill the threshold criterion
at lower fields may become effective close to B_{c2} (see Fig. 26)
and lead to a sharp peak in the pinning force density (PIPPARD
1969).

(c) Another reason for a peak close to B_{c2} could be a
change of the threshold criterion by flux lattice defects
(Section 2.2.2). The presence of an increasing number of such
defects at high flux densities may enable the flux lines to adjust to the pinning centers more easily than in an ideal lattice (CHANG et al. 1969). Although peaks due to mechanisms (b)
or (c) also occur at the same reduced flux density (b) for different temperatures, they are usually quite narrow and too
close to H_{c2} to be caused by mechanism (a).

4.4 Scaling Laws

Comparing the flux density dependence of the bulk pinning
force measured at different temperatures, one finds for many
samples a striking similarity of the shape of these curves.
This suggests plotting the results as a function of the reduced

flux density $b = B/B_{c2}$ so that they all coincide at $b = 1$. If one furthermore normalizes the $P_v(b)$ curves to their maximum values P_{max}, a single curve is obtained in most cases. Looking at Eqs. (4.13) and (4.16) this is not very astonishing since all variables in these relations can be expressed as functions of b (see Sections 2.2, 3.1, and 3.2) then assume the form

$$P_v = D \frac{B_{c2}(T)^n}{\kappa^m} f(b) \tag{4.26}$$

Here D, n, and m are material constants and f(b) is a function only of the reduced flux density b.

A scaling law of a form similar to (4.26) was first introduced by FIETZ and WEBB (1969) for a phenomenological description of experimental results on cold-worked Nb-based alloys. In the meantime it could be shown (see e.g. KRAMER 1973) that such a scaling law holds for a large variety of hard superconductors ranging from model systems such as Pb-In (COOTE et al. 1972) or Nb-Ta (ANTSBERGER and ULLMAIER 1974) to the commercial materials Nb-Ti (WOHLLEBEN 1973, see Fig. 30) and Nb_3Sn (HALLER and BELANGER 1971).

The recognition of scaling laws for the pinning force density is important for two reasons. The first is a practical one because one needs only to measure P_v, and thus f(b) at one temperature and one is then able to predict P_v for other temperatures simply by scaling the results by $B_{c2}(T)^n/\kappa^m$. The second benefit of plotting reduced P_v curves as a function of b is that one can sometimes obtain quite useful qualitative information about the pinning mechanism, particularly in cases when a scaling law is not obeyed. For instance, matching between flux line distance and pinning spacing will show a maximum in P_v at a fixed flux density, rather than reduced flux density and there will be no unique function f(b). Another example is the pinning by particles which are superconducting at low fields and temperatures but become normal conducting at fields and temperatures which are below the H_{c2} and T_c values of the matrix (see e.g. LIVINGSTON 1966).

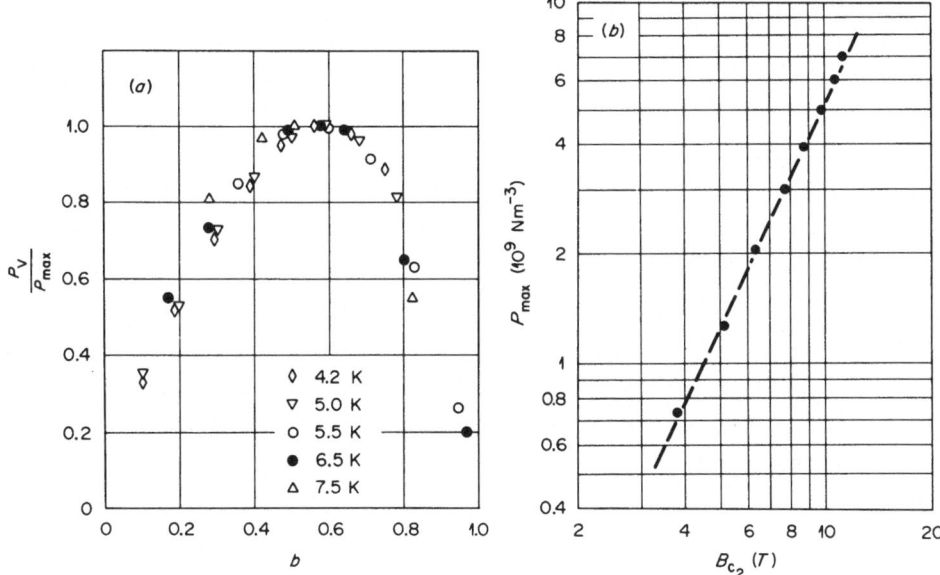

Fig.30 (a) Normalized pinning force density P_v/P_{max} as a function of the reduced flux density b for Nb-50 wt % Ti wire at different temperatures. (b) Log-log plot of the maximum pinning force density P_{max} vs the upper critical field B_{c2} which is a function of temperatur. The dashed line through the experimental points has a slope n of about 2 (after WOHLLEBEN 1973).

On the other hand, if a scaling law of the form (4.26) does hold, one may be fairly certain that a well defined pinning mechanism is operating over the whole field and temperature range and that the theoretical expressions given in Section 4.2 may be used for an analysis of the experimental results.

5. Some Experimental Results

Upon looking over the literature on hard superconductors one realizes a rather strict division of the experimental work which is unfortunately common to many fields of science and technology. On the one hand are the experiments on metallurgically simple "model" systems where one tries to correlate the measured bulk pinning force with one of the pinning mechanisms discussed in Chapter 3. On the other hand we find critical current measurements on materials in technological use. These measurements provide the necessary information needed for the design and construction of superconducting magnet coils. Combined efforts of physicists and metallurgists are now beginning to narrow the gap between these two groups, however, there is still a long way to travel towards a detailed understanding of pinning processes in complicated metallurgical systems. Therefore, although reluctantly, I find myself compelled to a similar division in this chapter: in the first part some typical results from samples containing only one kind of pinning centers with known distribution will be discussed whereas in the second part critical current data for several commercial hard superconductors are listed.

5.1 Experiments on Model Systems

5.1.1 Precipitates

Second phase precipitates are usually very strong pinning centers which can interact with the flux lattice via different mechanisms, depending on their size, shape, and composition. Probably the most thorough investigation of pinning by precipi-

tates was carried out for Bi-precipitates in the ε-phase of
eutectic Pb-Bi alloys (CAMPBELL and EVETTS 1966, CAMPBELL et al.
1968, COOTE et al. 1972). In this system the size of the normal
conducting Bi precipitates (2 to 10 μ) and the mean spacing
between them is orders of magnitude larger than the vortex spacings and we may expect that the rigidity of the vortex lattice
should not be an important factor in this case (see Sections
4.2.2 and 4.3). Indeed, COOTE et al. (1972) could show that the
measured pinning force density P_v is in close agreement with a
relation derived by assuming that each pinning center acts independently with its maximum interaction force K_o: since there
are $1/a_o$ flux lines pinned per unit length of precipitate normal to the vortices, P_v should be given by Eq. (4.18), i.e.

$$P_v = N_v A \frac{1}{a_o} \frac{K_o}{L_z} \qquad (5.1)$$

where $N_v A$ is the surface area of the precipitates per unit
volume and L_z is their length parallel to the flux lines.

Large normal conducting precipitates will pin predominant-
ly by means of the magnetic interaction (see Section 3.1.2).
Inserting Eq. (3.9) for K_o/L_z into Eq. (5.1) and expressing
M_{rev}, λ, and a_o in terms of B_{c2}, b, and κ yields

$$P_v = 0.6 \frac{N_v A B_{c2}^2}{\mu_o} \frac{1}{\kappa^3} b^{1/2} (1 - b) \qquad (5.2)$$

Fig. 31 shows that this equation describes the measured field
and temperature dependence very well and there is also an order of magnitude agreement between the theoretical and experimental P_v values (CAMPBELL and EVETTS 1972, p. 376).

For smaller precipitate sizes and distances the situation
becomes much more complicated because flux lattice rigidity
effects become important and the bulk pinning force will now
be given by expressions similar to Eqs. (4.13) or (4.16) rather
than by the above simple summation of individual interactions.
An example of pinning by precipitates for which Eq. (4.13) should

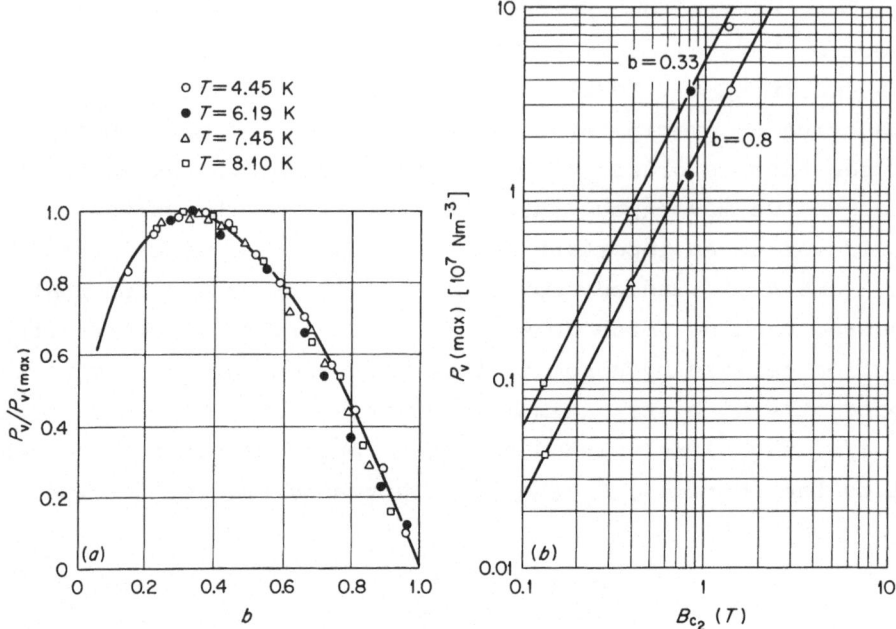

Fig.31 (a) Normalized pinning force density P_v/P_{vmax} as a function of b for large normal conducting Bi-precipitates in eutectic Pb-Bi. The experimental points for different temperatures fit the solid curve which is proportional to $b^{1/2}$ (1-b) very well. (b) Experimental variation of P_v with B_{c2} at different temperatures for reduced flux densitites of 0.3 and 0.8, respectively. The solid lines are of slope 2 (after COOTE et al. 1972, from CAMPBELL and EVETTS 1972).

be applicable was investigated by ANTESBERGER and ULLMAIER (1974) who measured P_v (B,T) in NbTa alloys ($\kappa \sim 5$) containing different concentrations of disk-shaped $(NbTa)_2$ N precipitates. These incoherent precipitates which have a diameter of about 1800 Å and a thickness of about 400 Å are normal conducting. By different aging treatments their concentration N_v could be varied from 4 to 50 x 10^{18} m^{-3}. Figure 32 shows that again a scaling law of the form (4.26) is valid and that the pinning force density exhibits a broad maximum around b = 0.5. In order to determine the interaction force K_0 we may proceed in the following way: transmission electron microscopy and metallography data give $N_v \bar{L}_y$ and magnetization measurements on reversible samples (i.e. those containing no precipitates) give the elastic constants C_{44} and C_{66} of the vortex lattice (see Section 2.2.1). Inserting these quantities and the measured P_v values into Eq.

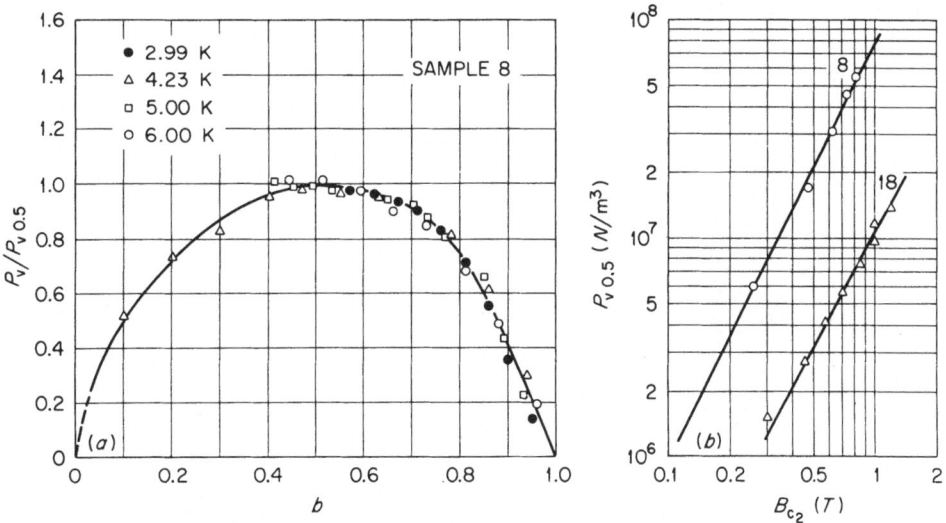

Fig.32 (a) $P_v/P_{v0.5}$ as a function of b for normal conducting $(NbTa)_2N$ precipitates in a Nb-Ta matrix. The function f(b) see Eq.(4.26) was found to be the same for all precipitate concentrations investigated. $P_{v0.5}$ is the maximum value of the pinning force density which is reached around b = 0.5 in this case. (b) Variation of $P_{v0.5}$ with B_{c2} (i.e., with temperature) for two samples containing different precipitate concentrations N_v. It was found that P_v is proportional to N_v (ANTESBERGER and ULLMAIER 1974).

(4.13) yields the maximum interaction force K_o which is shown in Figs. 33 and 34 as a function of flux density and temperature, respectively. In order to find which mechanisms are operating in the above system we may compare the data with theoretical predictions from the different interaction mechanisms discussed in Chapter 3. This procedure is simplified by disregarding a priori the interactions which certainly play no role in this case: there will be no pinning due to para- or ferro-magnetic particles and elastic interactions will be negligible since no indications of strain fields around the matrix-precipitate boun-dary could be detected in the electron microscope. From the two remaining mechanisms, the core interaction should dominate be-cause dividing Eq. (3.4) by Eq. (3.13) yields K_o (core)/ K_o (magn) = $\kappa/2 \ln \kappa \simeq 2$ for our case. We should therefore com-pare the experimental K_o values for low flux densities with Eq. (3.4). This is done in Fig. 34 which shows that the theore-tical H_{c2} (or T) dependence is in excellent agreement with the

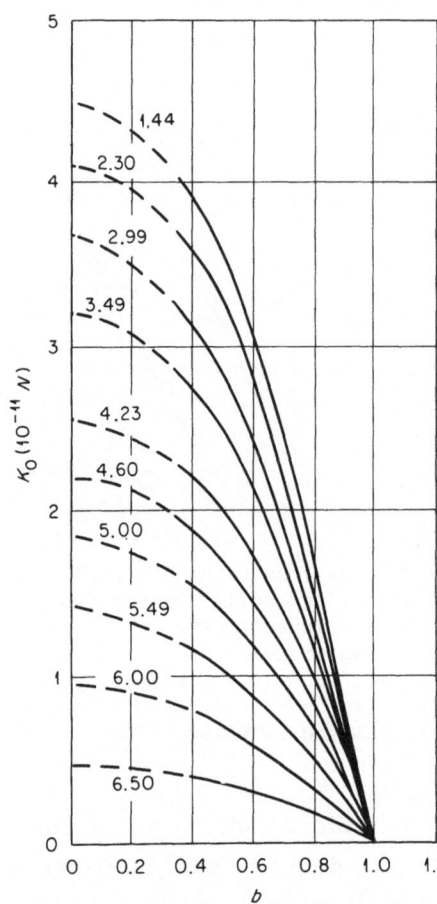

$K_O \, (10^{-11} \, N)$

1.44
2.30
2.99
3.49
4.23
4.60
5.00
5.49
6.00
6.50

b

Fig.33 Interaction force K_O as a function of the reduced flux density b for a Nb-Ta sample with $L_z \simeq 1050$ Å at different temperatures. The K_O values were obtained from Eq.(4.13) using the experimental values for P_V, N_V, $\bar{L}_y$, C_{44} and C_{66} (ANTESBERGER and ULLMAIER 1974).

experiments and that the order of magnitude of the absolute values of K_O is also correctly predicted by Eq. (3.4).

These findings are a strong indication that the core interaction is indeed the dominant pinning mechanism in this model system. We are now able to estimate the interaction energy between a flux line and a $(NbTa)_2 N$ precipitate by multiplying K_O with the interaction length $d \simeq \xi$. For $b \to O$ and $T = 4.2$ K we obtain $U \simeq 2.5$ eV. This value is in order-of-magnitude agreement with the barrier height U_O obtained from flux creep experiments performed on the same samples (see Section 6.1).

Additional results which support the above considerations were obtained from single crystalline nitrogen-doped Nb samples of different orientation (ANTESBERGER and ULLMAIER 1975). Because of the disk-like shape of the Nb_2N precipitates which lie

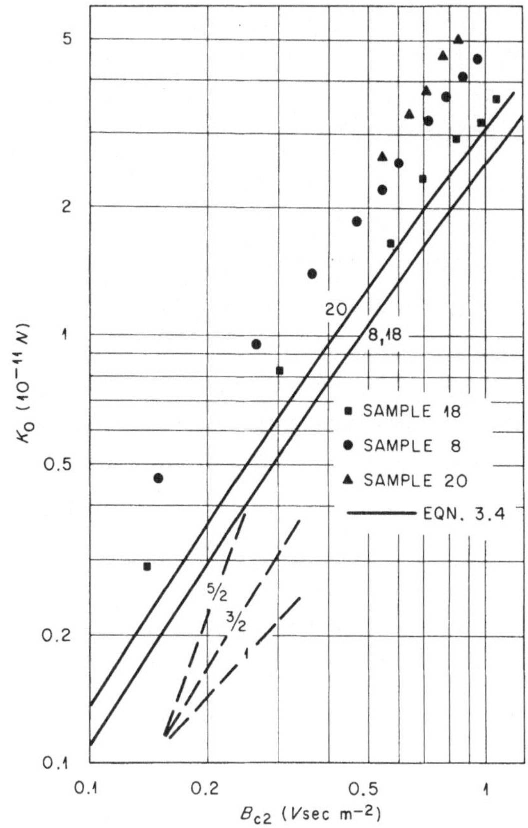

Fig.34 Comparison of the temperature (or B_{c2}) dependence of theoretical and experimental K_O values for $b \to 0$. The theoretical values (full lines) were obtained by inserting experimental numbers for H_{c2}, κ, and $\bar{L}_z$ into Eq.(3.4) which contains no adjustable parameter. The fact that the experimental data are always higher may be attributed to the magnetic interaction which is not included in Eq. (3.4) and which has practically the same temperature dependence as the core interaction.

on {100}-planes, the bulk pinning force should strongly depend on the angle between field direction and crystal axis. If a <100> -axis is parallel to $\underline{B}$, 2/3 of the precipitates will have its large dimension parallel to the vortices (resulting in a large $\bar{L}_z$). This is not the case for a sample with <111> ∥ $\underline{B}$ where $\bar{L}_z$ is given by t/sin 35° with t being the small thickness of the precipitates. A more detailed geometrical analysis shows that the ratio P_V <100> /P_V <111> should be 9.0 which agrees well with the experimental ratio of 8 ± 1.5.

This agreement is a further indication, that a summation procedure which takes lattice rigidity effects into account ($P_V \propto K_O^2$) must be used in this case. If the pinning centers would act independently ($P_V \propto K_O$), the theoretical ratio P_V <100>/ P_V <111> would only be 3, i.e. much too small to account for the observed ratio of around 8.

As a third example of pinning by precipitates some results obtained in Pb-Na alloys containing rather small Pb₃Na precipitates (L < ξ) may be briefly described (FREYHARDT 1971a). The interpretation of these data in terms of basic pinning interactions is difficult for two reasons. Firstly, the mean distance ℓ between precipitates ranges between 200 and 2000 Å which is smaller or at most equal to the vortex spacings. That is to say, the pinning array is far from being dilute (see Section 4.3) and it is uncertain how accurate an evaluation with Labusch's theory will be. Another problem is that at least two pinning mechanisms are expected to operate simultaneously: a dielastic interaction caused by the lattice parameter mismatch and a core interaction due to different superconducting properties of precipitates and matrix. Despite these complications, Freyhardt was able to estimate K_o (see Fig. 35) by applying Eq. (4.13) and the values obtained seem to agree with the theoretical estimates given by Eqs. (3.2) and (3.20).

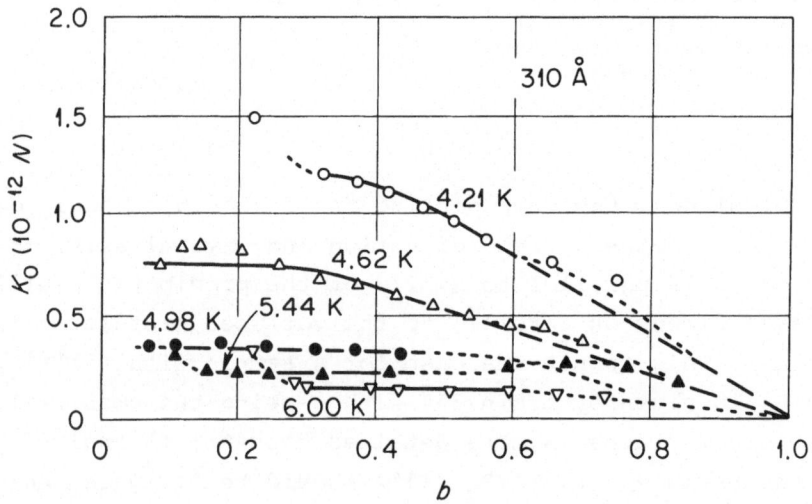

Fig.35 Interaction force K_o as a function of the reduced flux density b between a Pb₃Na particle of 310 Å diameter and a flux line in PbNa (after FREYHARDT 1971a).

Investigations of pinning by precipitates in Nb-Ti alloys have been carried out by BAKER and SUTTON (1969) and HILLMANN

and HAUCK (1972). Although these results are very important
for a better understanding of the pinning processes in the
widespread technological material Nb-Ti, the data are not sufficient to attempt a quantitative analysis.

5.1.2 Dislocations

There are a large number of pinning force measurements on
samples containing dislocations, mainly because it is assumed
that the dislocation structure is the preponderant reason for
the high critical currents in commercial Nb-Ti alloys. However,
in spite of the extensive experimental data available (for a
list of references see, e.g., VAN GURP and VAN OOIJEN 1966a,
and DEW-HUGHES and WITCOMB 1972), our understanding of the interaction between flux lines and dislocation structures is still
poor for various reasons. Even for dilute and fairly uniform
dislocation distributions (GOOD and KRAMER 1970, COOTE 1970,
FREYHARDT 1969, 1971b) it is very unlikely that vortices will
be pinned by individual dislocations since the most optimistic
estimates of dK/dx for this case give values which are more
than two orders of magnitude smaller than the pinning threshold
(see Section 4.1 and 4.2.3). Indeed, an analysis of experimental results from torsion-deformed Nb crystals by FREYHARDT
(1971b) where an individual vortex-dislocation interaction was
assumed, yields K_o values which are ten times larger than the
theoretical estimate Eq. (3.19) for the dielastic interaction
for screw dislocations. Freyhardt therefore concludes that the
pinning is caused by regions of high dislocation density which
should contain about 100 individual dislocations. Qualitatively this view is supported by electron micrographs which show a
non-statistical dislocation structure.

A similar conclusion was reached by COOTE (1970) who studied pinning effects in Nb-Ta alloys deformed in tension. These
alloys maintain a nearly uniform dislocation structure to much
higher deformations than pure Nb and have the advantage of a
higher Ginzburg-Landau parameter κ which makes a theoretical
analysis easier. Coote was able to measure the field dependence
of P_v down to very low flux densities and could show that $P_v(b)$

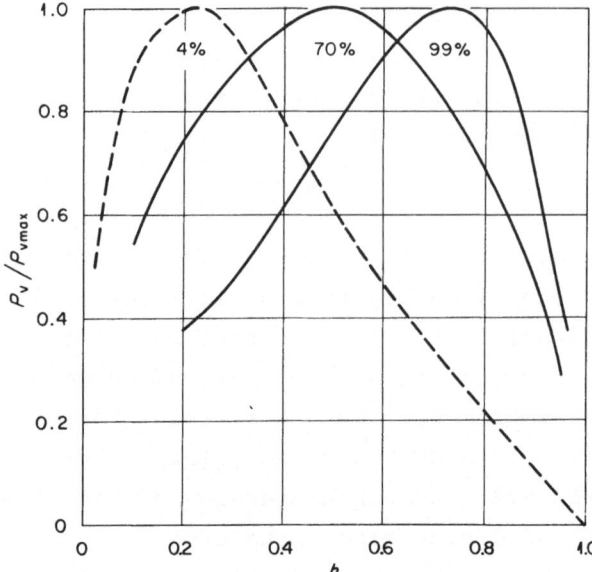

Fig.36 The change in the position of the maximum pinning force density for different degrees of deformation in Nb-50 at.% Ta alloys. The dotted curve corresponds to a nearly uniform dislocation structure. At 99 % deformation the cell structure was fully developed (after COOTE 1970, from CAMPBELL and EVETTS 1972).

exhibits a maximum around b = 0.25 (see dotted curve in Fig. 36) and scales with B_{c2}^3. In order to explain this behavior, CAMPBELL and EVETTS (1972, p. 385) preposed that the pinning is mainly caused by groups of dislocations which are just strong enough to fulfill the threshold criterion. Using a temperature and field dependence of B_{c2} (1-b) for the elastic interactions they then find

$$P_v \propto (2L_y N_v) \ B_{c2}^{3/2} \ \kappa \ b^{1/2} \ (1 - b) \qquad (5.3)$$

which is close to the observed field dependence (Eq. (5.3) which has its maximum at b = 0.33 is only valid at high flux densities; if the correct low field expressions for the elastic constants are used, the maximum is near b = 0.25). However, in order to explain the experimental B_{c2}^3 dependence one must assume that the product $2L_y N_v$ is temperature dependent (e.g., $L_y \propto \xi$ and $N_v \propto 1/\xi^2$) but at present there is no evidence to support such a hypothesis.

Campbell and Evetts also discuss upper limits for the pinning force density and the concentration of pinning points which can still be analyzed by the dilute approximation in Labusch's theory (see Eq. (4.13)). For materials with parameters between those of Nb and Nb-Ta they find $P \stackrel{\sim}{<} 2 \times 10^5$ Nm^{-3} and

84

$N_v \overset{\sim}{<} 3 \times 10^{12} P_v (m^{-3})$ where P_v is in Nm^{-3}. Freyhardt's specimens with P_v values around 2×10^6 Nm^{-3} at $b = 0.5$ and $T = 4.2$ K and most of the results of other investigations fall outside this limit.

The situation is even more difficult to survey for the case of nonuniform dislocation structures in heavily deformed metals because pinning can now occur through more than one in- teraction mechanism. Besides the dielastic and parelastic inter- action considered above, dislocation cell walls in which H_{c2} and κ are increased can lead to magnetic interactions (NARLIKAR and DEW-HUGHES 1964) and in anisotropic materials pinning will also occur through ΔM differences across subgrains or cell boun- daries. Another difficulty comes from surface layers (thickness ~ 10 μ) which, in many cases, have much higher critical current densities than the bulk material (COOTE 1970, ZERWECK 1973). This can lead to misinterpretations of magnetization curves and some A.C. measurements (see Chapter 7). Further complications arise when annealing treatments are involved between subsequent deformations as is the case during the production of commercial NbTi wires. Besides changing the defect structure of the alloy in a complex way, such annealing treatments can cause second phase precipitates which are nucleated in sub-bands (HILLMANN and HAUCK 1972).

A systematic investigation of the field dependence of P_v as a function of the degree of deformation was carried out by COOTE (1970 in Nb-Ta alloys. With increasing deformation he finds a shift of the maximum of the $P_v(b)$ curves towards higher reduced flux densities until the maximum occurs around $b = 0.75$ for a fully developed cell structure (Fig. 36). The absolute P_v values are then several orders of magnitude larger than for the lightly deformed samples containing nearly uniformly distribu- ted dislocations.

Measurements of pinning forces as a function of B and T in heavily cold-worked Nb-Ti alloys were reported by FIETZ and WEBB (1969) and HAMPSHIRE and TAYLOR (1972). Both find similar field dependences (Fig. 37) and a scaling of P_v proportional to B_{c2}^n with n ranging between 2 and 3 (Fig. 38). Their approa-

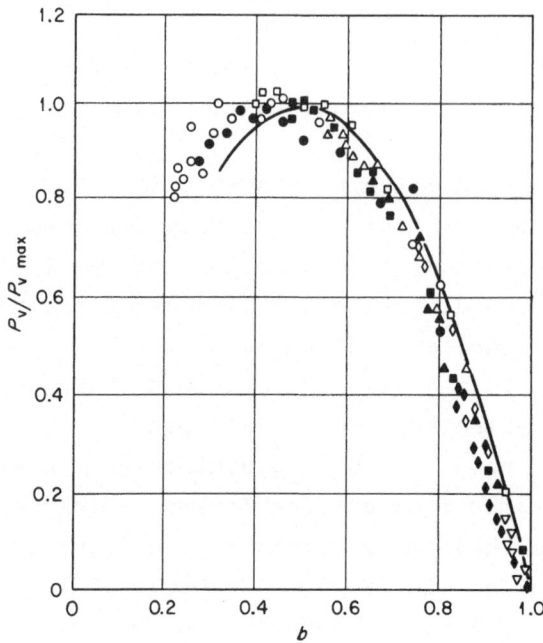

Fig.37 Normalized pinning force density as a function of reduced flux density b in a commercial Nb-Ti alloy at different temperatures. The solid curve shows the b dependence of Eq.(5.5).
At b = 0.5 and T = 5 K, P_{vmax} = 4.2 x 10^9 Nm^{-3} (after HAMPSHIRE and TAYLOR 1972).

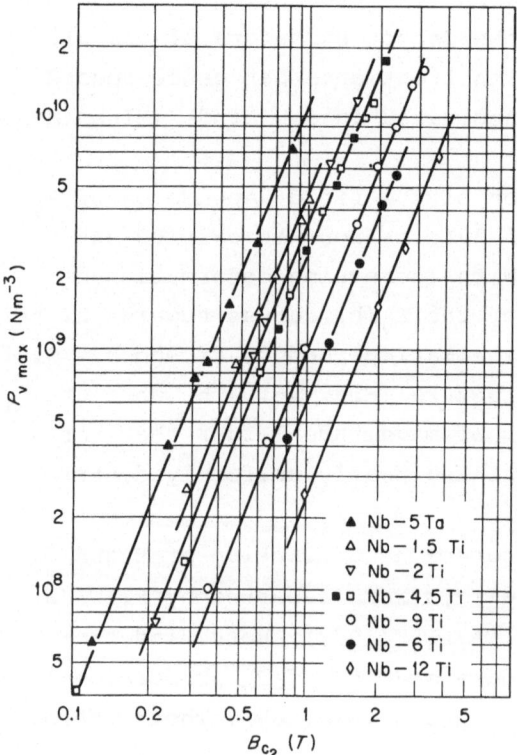

▲ Nb − 5 Ta
△ Nb − 1.5 Ti
▽ Nb − 2 Ti
■ ▫ Nb − 4.5 Ti
○ Nb − 9 Ti
● Nb − 6 Ti
◇ Nb − 12 Ti

Fig.38 Temperature dependence of the maximum pinning force density in deformed NbTi alloys. The solid lines have a slope 5/2 in agreement with Eq.(5.4) (after FIETZ and WEBB 1969).

86

ches to interpret the experimental results in terms of basic pinning mechanisms, however, are quite different. Fietz and Webb use Labusch's theory and assume that groups of dislocations interact with the flux lattice through the dielastic mechanism and that $K_o \propto H_{c2}(T)^{3/2}$. Combining the field, temperature, and K dependence of all the factors in Eq. (4.13) they find

$$P_v = D \, \frac{B_{c2}^{5/2}}{\kappa^2} \, b \qquad\qquad \text{for } b \to 0$$

$$P_v = D' \, \frac{B_{c2}^{5/2}}{\kappa^3} \, b \, (1 - b) \qquad \text{for } b \to 1 \qquad\qquad (5.4)$$

(where D and D' are material constants) which is in reasonable agreement with the experimental results.

Hampshire and Taylor attempted to explain their results by considering the free energy change of a vortex when entering a region of higher κ. Neglecting the rigidity of the vortex lattice they then simply sum up all individual interaction forces per unit volume and obtain

$$P_v = \frac{\sqrt{3}\ \omega}{4L\ \mu_o\ \beta_A}\ \frac{B_{c2}^2\ \Delta\kappa}{\kappa^3}\ b\ (1 - b) \qquad\qquad (5.5)$$

where $\Delta\kappa$ is the difference between the Ginzburg-Landau parameters in cells and cell walls, L is the cell size and ω is the proportion of cell wall oriented for pinning. Although based on entirely different assumptions, Eqs. (5.4) and (5.5) have a rather similar field and temperature dependence which again shows how difficult it is to identify the correct pinning mechanism from macroscopic measurements.

The importance of lattice rigidity effects for the bulk pinning force in high κ-superconductors has also been questioned by DEW-HUGHES (1974a,b). His considerations are mainly based on an extensive study of the effect of dislocation tangles on the superconducting properties of deformed Mo-Re alloys (DEW HUGHES and WITCOMB, 1972). This work contains quantitative information about the differences $\Delta\kappa$ and ΔH_{c2} between cells and cell walls

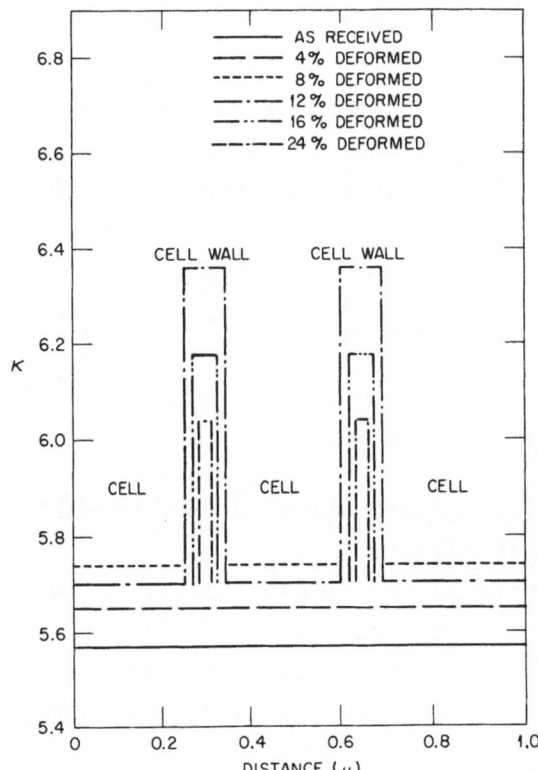

Fig.39 Variation of Ginzburg-Landau parameter κ with distance through Mo-39 at. % Re alloys subjected to increasing deformation by rolling (after DEW-HUGHES and WITCOMB 1972).

(Fig. 39) which confirms the importance of magnetic interactions in materials with nonuniform dislocation structures. In order to correlate pinning force measurements on the same samples with the data on $\Delta\kappa$ and ΔH_{c2}, Dew-Hughes and Witcomb assume that P_v is given by Eq. (5.2) which was derived for widely spaced interfaces between regions with magnetization M_{rev} and $M_{rev} + \Delta M_{rev}$, respectively (see Section 3.1.2 and Fig. 29). Using Eq. (8.25) for M_{rev} as a function of B, κ, and H_{c2} they obtain

$$\Delta M_{rev}(B) = \frac{B_{c2}-2B}{2\mu_o \beta_A} \frac{\Delta\kappa}{\kappa^3} \tag{5.6}$$

which, inserted into Eq. (5.2) gives

$$P_v = \frac{L^2}{\ell^3} \frac{(2\pi)^{1/2}}{2\mu_o \beta_A} B_{c2}^2 \frac{\Delta\kappa}{\kappa^4} b^{1/2} |(1-2b)| \tag{5.7}$$

where L is the cell diameter and ℓ is the center-to-center distance between adjacent cells. Inserting experimental values for L, ℓ, B_{c2}, κ, and $\Delta\kappa$ into Eq. (5.7) yields P_v values which are higher than the measured values by a factor of 6 to 15, depending on $\Delta\kappa$. Dew-Hughes and Witcomb attribute this discrepancy to the small cell wall thickness ($<2\lambda$) which prevents the actual value of ΔM_{rev} from reaching the bulk value calculated by Eq. (5.6) and P_v will be reduced. It is, however, more likely that the relatively small distance between pinning centers ($\ell \simeq 0.3 \mu \simeq 5 a_o$) is not sufficient to have every pinning center acting with its maximum force which was required in the derivation of Eq. (5.2). Actually, the pinning centers are so close that the array is "concentrated" (see Section 4.3) and we might expect that $P_v \propto \Delta M_{rev}^n$ where n is somewhere between 1 and 2 (see Eq. (4.25)).

The above examples show that at present both the available experimental data and the state of the theory are insufficient to attempt a quantitative description of pinning by dislocation structures and to clarify the importance of different pinning mechanisms, flux lattice rigidity, etc.

5.1.3 Grain boundaries and other defects

For a given field direction, two adjacent grains in an anisotropic superconductor generally have different superconducting properties (e.g., H_{c2} of $Nb_{85}Ta_{15}$ at 2.4 K varies between 0.694 and 0.704 T, depending on the orientation of the crystal with respect to the magnetic field (YAMAMOTO et al. 1974). Since this situation is similar to that of boundaries between a precipitate ard a matrix we may expect that grain boundaries in anisotropic materials will pin vortices through the magnetic interaction. This view is supported by the field and temperature dependence of P_v in Nb_3Sn (Fig. 40) which is very similar to that for large second-phase precipitates (Fig. 31), i.e., P_{vmax} occurs at b = 0.33 and scales with B_{c2}^2. In order to estimate the order of magnitude of grain boundary pinning we calculate the change in the core energy of a vortex moving from one grain with a thermodynamic critical field H_c to an adjacent grain with $H_c' =$

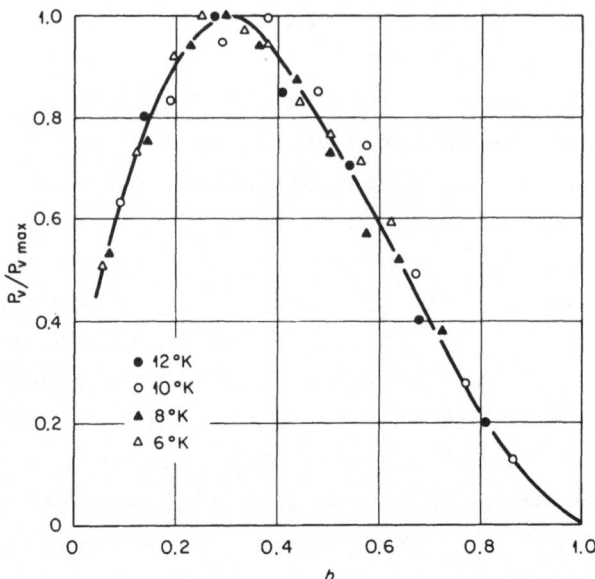

Fig.40 Normalized pinning force density as a function of reduced flux density b in Nb_3Sn tape at different temperatures (from KRAMER 1973 after data from HALLER and BELANGER 1971).

$H_c + \Delta H_c$. Assuming $\Delta H_c/H_c = 0.05$, Eq. (3.3) then yields a pinning force per unit length $K_o/L_z \simeq 10^{-4}$ Nm^{-1} for Nb_3Sn ($B_{c2} = 23$ T, $\kappa = 35$).

There is now clear evidence that grain boundaries are the dominant pinning sites in some β-tungsten (A-15) compounds. Measurements of HANAK and ENSTRÖM (1966) in vapor deposited Nb_3Sn and NEMBACH and TACHIKAWA (1969) in V_3Ga show a correlation between grain size and pinning force over a wide range. A striking qualitative demonstration of the importance of grain boundary pinning are measurements on V_3Si single crystals by PULVER (1972) who found extremely low critical current densities of around 10^4 Am^{-2} as compared to 10^9 Am^{-2} for the polycrystalline material. Recently SCANLAN (1974) measured $P_v(B)$ for Nb_3Sn samples formed by solid state reaction. With this process one can obtain different grain sizes by choosing different reaction temperatures without using stabilizing additives like Zr. The material is, therefore, free of precipitates which simplifies an interpretation of the experimental results. Fig. 41 shows the maximum bulk pinning force as a function of the in-

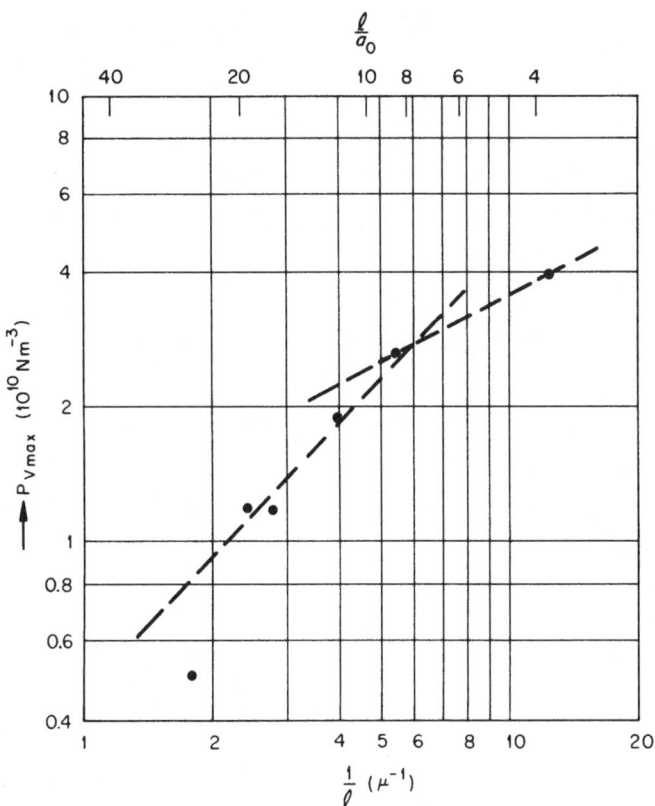

Fig.41 Log-log plot of the maximum bulk pinning force P_{vmax} vs the inverse grain size ℓ^{-1} for Nb_3Sn formed by solid state reaction (SCANLAN 1974). For grain sizes down to about 8 times the flux line spacing a_0, P_v increases linearly with ℓ^{-1}. For smaller grains a slower increase is observed (the dashed lines have a slope of 1 and 1/2, respectively).

verse grain size ℓ. Qualitatively this plot is similar to Fig. 27 and indicates that grains with diameter $\ell > 2000 \overset{o}{A} \simeq 10 \, a_0$ represent a dilute pinning array ($P_v \propto \ell^{-1}$) whereas smaller grains show the behavior of concentrated array ($P_v \propto \ell^{-1/2}$). Inserting the K_0/L_z value of $10^{-4} \, Nm^{-1}$ calculated above for Nb_3Sn into Eq. (4.18) yields $P_v = 2.10^{10} \, Nm^{-3}$ for $\ell = 2500 \overset{o}{A}$ and $B = 5T$ which is very close to the experimental value (Fig. 41). Since Eq. (4.18) is valid for independently acting planes (see Section 4.2.2) this agreement might be taken as an indication that flux lattice rigidity effects can be neglected in this case. However, this assumption would lead to a field dependence of

$P_v \propto b^{3/2}$ (1 - b) (DEW-HUGHES 1974a) which is not observed experimentally. Furthermore, the $\Delta H_c/H_c$ value of 5 % was chosen rather arbitrarily. Therefore, a quantitative analysis must be postponed until data of the orientation-dependence of H_{c2}, κ, and H_c are available (such data can be obtained from magnetization and tunneling measurements on single crystals as e.g., those done for Nb by FARRELL et al. (1968) and MAC VICAR and ROSE (1968)).

Pinning by grain boundaries has also been observed in cold-rolled Nb (VAN GURP and VAN OOIJEN 1966b), and in Nb containing small amounts of Y or Gd (KOCH and KROEGER 1974).

Other defects which can act as pinning centers are interstitial and vacancy agglomerates produced by irradiation with high energy neutrons or charged particles. The influence of irradiation on superconducting properties is a rather extensively studied field for two reasons. First, the irradiation produces relatively simple defects and their concentration and structure can be changed in one and the same sample by varying the irradiation dose and the annealing temperature, respectively (for a survey on radiation induced defects in metals see, e.g., SEEGER et al. 1970). Second, radiation effects could affect the behavior of superconducting magnets in fusion reactors, a problem which is the motivation for many irradiation experiments on commercial materials.

There is only one measurement of pinning by statistically distributed single Frenkel pairs produced by low temperature electron irradiation in Nb (ULLMAIER et al. 1970). As expected, the observed pinning force density is very small and saturates with increasing defect density. The results can be interpreted by considering the statistical fluctuations in the defect density in volumes ξ^3. Regarding these volumes as pinning centers, a dielastic interaction in Eq. (4.13) seems to describe the observed order of magnitude of P_v and its field dependence fairly well.

Much stronger pinning effects are caused by nonuniform point defect arrangements (defect cascades produced by low temperature reactor irradiation or vacancy and dislocation loops formed during neutron irradiation above room temperature). This

92

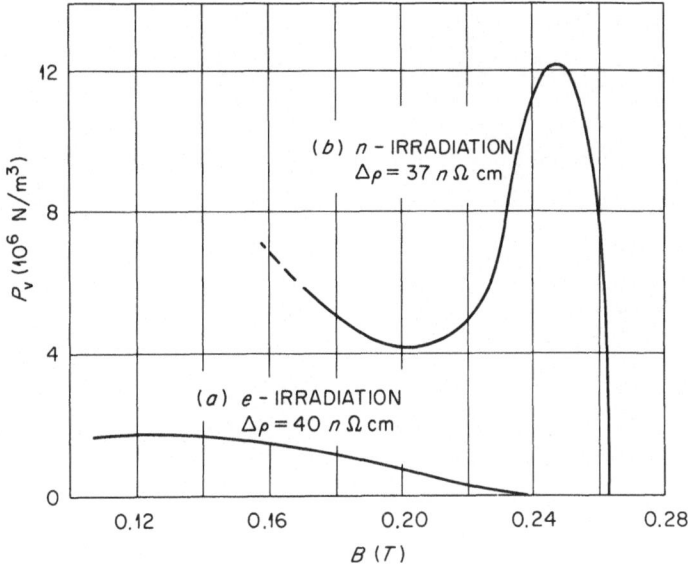

Fig.42 Pinning force density in two Nb samples containing about the same mean density of Frenkel pairs which are distributed (a) homogenously and (b) in displacement cascades. The neutron data (b) are from BERNDT et al.(1968).

is shown in Fig. 42 where the field dependences of P_V in two Nb samples containing the same mean defect density but different defect structures are compared (ULLMAIER 1970). There are a large number of investigations on the effect of fast neutron and charged particle irradiations on the critical current density in both pure metals and commercial alloys and compounds (for a list of references see, e.g., ULLMAIER 1973). In general one finds a drastic increase of P_V in materials which showed little or no pinning before irradiation, and a slight decrease in P_V in alloys with initially very high P_V values.[+] Several qualitative models have been proposed to explain these effects but only one quantitative analysis has been attempted so far (VAN DER KLEIN et al. 1974).

[+] The spectacular drop of P_V (and T_c) in Nb_3Sn at fluences beyond 10^{18} neutrons/cm^2 (PARKIN and SCHWEITZER 1973 and SWEEDLER et al. 1974) is caused by disordering effects and is not directly connected with a change in the pinning structure.

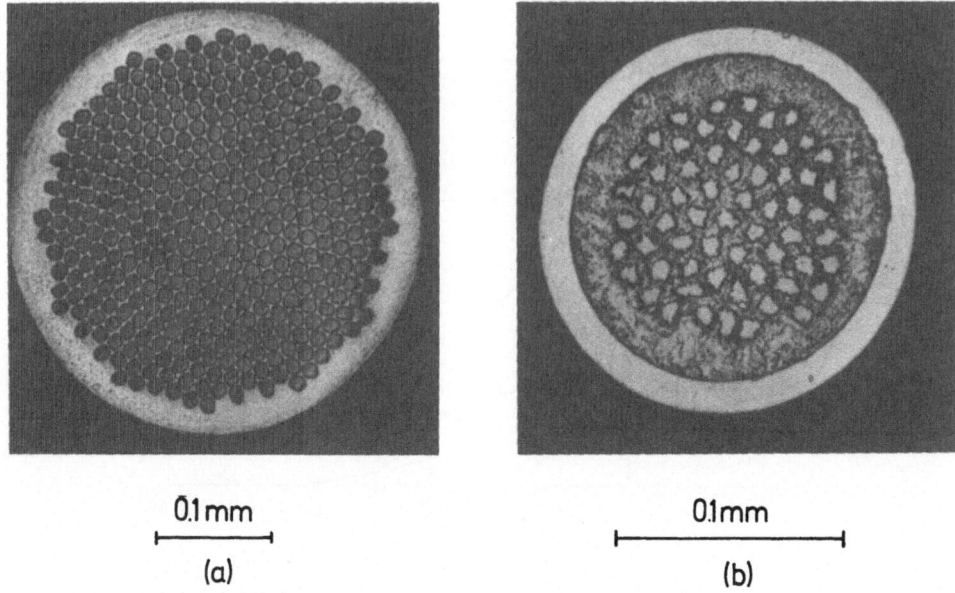

0.1 mm

(a)

0.1mm

(b)

Fig.43 Cross-sections of (a) 361 strand Nb-Ti multifilament conductor as drawn in copper matrix (Magnetic Corporation of America) and (b) 61 strand Nb_3Sn multifilament conductor as drawn in copper matrix (Intermagnetics General Corporation).

Interesting systems for pinning studies are metals which contain voids (diameter between 100 and 1000 Å) produced at 0.3 - 0.5 of the melting temperature by high fluence irradiation (10^{21}-10^{23} neutrons/cm^2). Some preliminary results (FREY-HARDT 1974), show that such voids which sometimes form an ordered array (a void superlattice) should be an excellent model system for testing pinning theories, the influence of concentrated arrays, matching effects, etc.

5.2 Short Sample Characteristics of Commercial Materials

At present the technical application of hard superconductors is mainly confined to magnet coils used for a large variety of purposes ranging from small research magnets to coils for electrical power generators and motors, plasma confinement, high energy physics, etc. For magnetic fields up to about 8T the ductile alloy Nb-Ti has almost completely replaced Nb-Zr which was used in the early sixties. Today several manufacturers supply multifilament Nb-Ti/Cu composite wires (Fig. 43a) with excellent

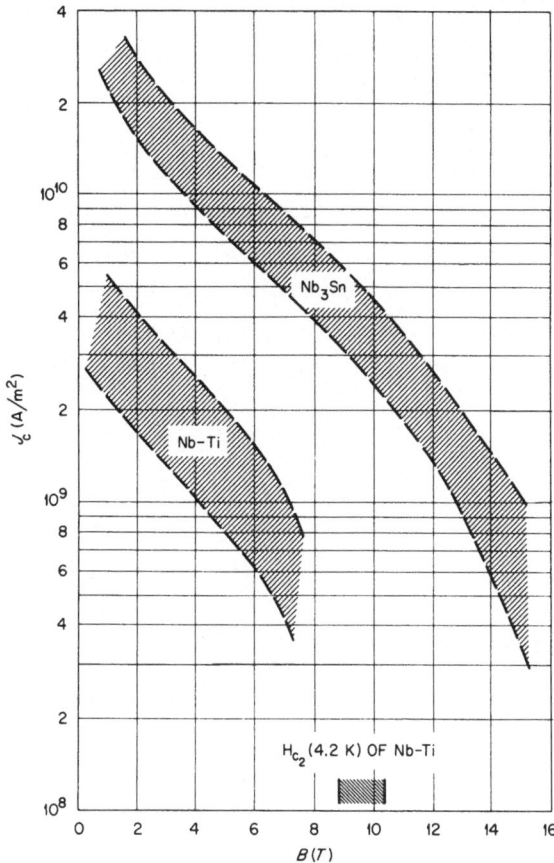

Fig.44 Range of critical current densities as a function of transverse
applied field for commercial Nb-Ti and Nb$_3$Sn material at 4.2 K.

electrical, thermal, and mechanical properties. These conductors are very versatile and can be tailored to meet the requirements of many applications. However, for fields above 8T the
critical current density of Nb-Ti becomes very small (Fig. 44)
and the compounds Nb$_3$Sn or V$_3$Ga which have much higher upper
critical fields must be used. With these materials magnets with
a central field up to 16.5T have been built and it is expected
that fields above 18T will be reached in the near future (ROS-
NER 1974).

Although the A-15 compounds themselves are brittle, modern
composit Nb$_3$Sn and V$_3$Ga tapes are rather flexible and can with-

stand tensile stresses up to several hundred MN/m². Very recently, several attempts to produce multifilament Nb₃Sn and V₃Ga
conductors in Cu-matrix (SUENAGA and SAMPSON 1971, TACHIKAWA
1972) provided encouraging results. It is almost certain that
these materials (Fig. 43b) will be available by early 1975 in
quantities sufficient for small research magnets and larger
prototype coils (FIETZ and ROSNER 1974, GREGORY et al. 1974).

The range of critical current densities achieved in commercial Nb-Ti and Nb₃Sn conductors as a function of an applied
transverse field is shown in Fig. 44. Although this diagram is

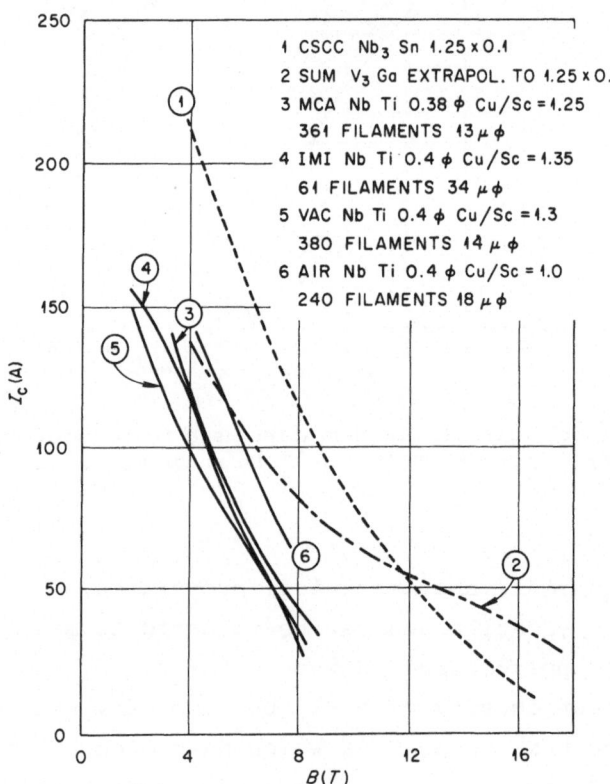

Fig.45 Some typical short sample characteristics of conductors used for
small research magnets. All conductors have about the same copper-to-
superconductor ratio (Cu/Sc) and the same overall cross section (CSCC =
Canada Superconductor and Cryogenics Company Ltd., Canada; SUM = Sumitomo
Electric Industries Ltd., Japan; MCA " Magnetic Corporation of America,USA;
IMI = Imperial Metals Industries, England; VAC = Vacuumschmelze GmbH,
Germany; AIR = Airco Kryoconductor, USA).

96

interesting for an analysis of basic pinning processes, it is
not very useful for a magnet designer because these materials
are always embedded in a suitable matrix, usually high conductivity copper in order to prevent instabilities (see Section 6.3).
It is therefore the total current carried by a conductor of given size which is relevant for the magnet design rather than the
critical current density in the superconductor alone. A plot of
the maximum total current I_c vs the applied transverse field H
is called the short sample characteristic of the conductor (in
most applications the field direction is perpendicular to the
conductor, i.e., to the current). Examples of short sample characteristics for some commercial conductors having about the
same cross section of 0.125 mm^2 are shown in Fig. 45. If no degradation occurs (see Section 6.3) the knowledge of the short
sample characteristic immediately permits one to determine the
maximum obtainable center field H_o of a solenoid of given size,
shape, and number of turns. In Fig. 46 the line $\overline{OP}$ shows the

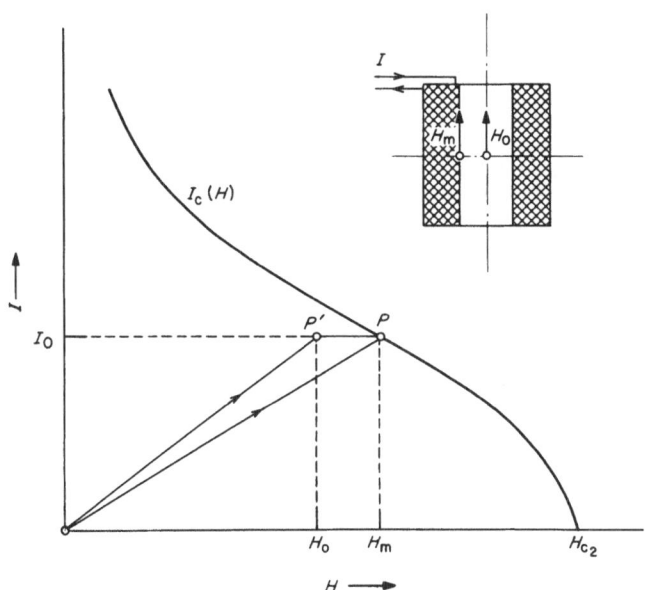

Fig.46   Determination of the maximum attainable center field H_o for a coil
of given dimension and wound of a superconductor with a short sample
characteristic $I_c(H)$. H_m is the maximum field at the winding and I_o is the
maximum operating current.

proportionality between the coil current I and the maximum magnetic field $H_m = k_m I$ which usually occurs in the innermost winding at the center plane of the solenoid. The maximum coil current I_o is reached when this line meets the short sample characteristic (point P). This current produces a central field H_o which is given by the intersect of the horizontal line $\overline{I_o P}$ with the line $\overline{OP'}$ given by $H_o = k_o I$. The constants k_m and k_o depend on the ratios of coil length and outer diameter to the inner diameter and are available in tabulated form (see e.g., MONTGOMERY 1969). For more complicated coil shapes (split magnets, quadrupols, saddle magnets, toroidal coils, etc.) the calculation of design parameters like H_m, H_o, I_o, stresses in the windings, stored magnetic energy, etc. is sometimes very difficult and elaborate computer programs must be employed. Readers who are interested in the challenging engineering problems arising in the technical application of superconductors in magnets and other fields are referred to the proceedings of the biannual Applied Superconductivity Conferences and the annual International Cryogenic Engineering Conferences. Short reviews of the physical principles of superconducting devices can also be found in the article by NEWHOUSE (1969), in BUCKEL's book on superconductivity (1972), and in the Proceedings of the IEEE, Jan. 1973, Vol. 61, No. 1.

6. Dissipative Effects

Heretofore we have assumed that the flux distribution in a hard superconductor at fixed external fields and currents is constant in time and is given by the critical state Eqs. (2.20) or (2.26). For most practical purposes this critical state concept indeed provides a sufficiently accurate description of the behavior of a hard superconductor. Very sensitive flux measurements, however, reveal a very slow motion of flux at current densities close to the critical current density J_c. This flux creep was first observed by KIM et al. (1962) and was attributed to thermally activated jumps of so-called flux bundles (ANDERSON 1962). If the current density (or flux density gradient) is increased above its critical value[+]) the creep-like flux motion changes into a viscous flow of flux lines (KIM et al. 1963, 1965). It is observed that flux motion is always accompanied by a macroscopic electric field $\underline{E}$ given by

$$-\underline{E} = \underline{v}_L \times \underline{B} \tag{6.1}$$

where $\underline{v}_L$ is the average flux line velocity. The corresponding power density input $\underline{E} \cdot \underline{J}$ which is required to maintain a steady flow of flux lines is dissipated in the specimen. When this power dissipation exceeds a certain limit, a thermomagnetic instability builds up and superconductivity breaks down completely (flux jump). In the following I shall briefly discuss flux creep, flux flow and flux jumps (see Fig. 47) and list some

---

[+] Since in principle flux creep continues as long as a flux density gradient exists, there is no precise "critical state". The critical flux density gradients observed experimentally thus represent only quasi-stationary states at which the creep rate has fallen below a practically observable limit.

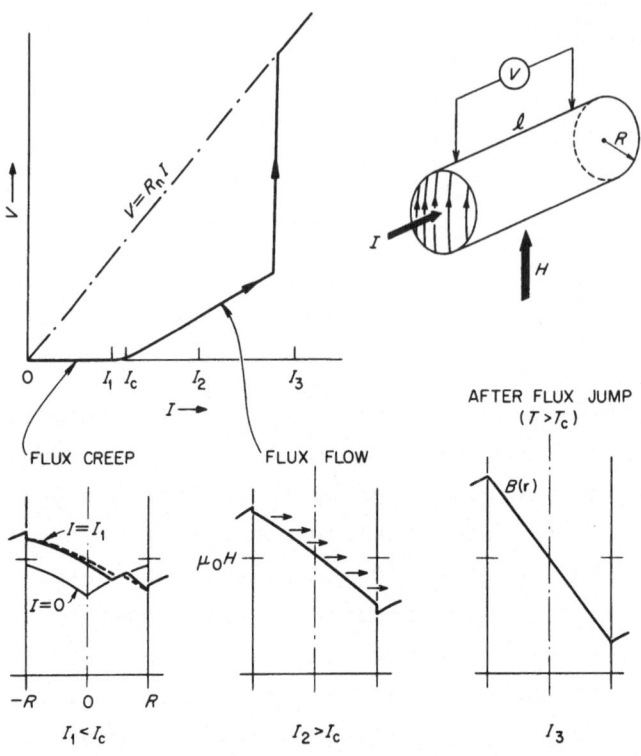

Fig.47 Voltage-current characteristic of a hard-superconducting wire in a transverse magnetic field H. For small currents I_1 below the critical current I_c, flux lines can move by thermally activated jumps. This flux creep produces only extremely small voltages V. For currents $I_2 > I_c$ the pinning forces are unable to support the Lorentz force $\underline{B} \times \underline{J} > \underline{P}_V$ and flux lines are continuously driven across the sample (flux flow). Such flux movement dissipates energy which finally drives the sample into the normal state (flux jump). At the bottom of the figure the field distributions for each of the three regimes are shown schematically.

methods which can be applied to prevent catastrophic instabilities in materials for technological use (stabilization).

6.1 Flux Creep

As first pointed out by ANDERSON (1962), at non-zero temperature flux motion is possible with the help of thermal activation even if $P_D < P_V$. Thus after an applied field (or current) change, vortices continue to move thereby relaxing the driving force P_D (case $I_1 < I_c$ in Fig. 47). Because of the strong coupling of vortices to each other, one thermally activated jump always involves several vortices simultaneously. According to Anderson's theory the rate ν_c at which such "flux bundles" jump over pinning barriers is given by

$$\nu_c = \nu_o \, e^{-U_{eff}/k_B T} \tag{6.2}$$

where ν_o is an attempt frequency and U_{eff} is an effective activation energy. If the driving force P_D is close to the pinning force P_v, U_{eff} is much smaller than the pinning potential U_o since the presence of a driving force strongly assists the motion of vortices past their pinning barriers. As the simplest

form with the correct physical features, Anderson and Kim used the linear relation

$$U_{eff} = U_o - P_D V_B X \tag{6.3}$$

where V_B is the activation (or bundle) volume and X is the width of the energy barrier. In order to analyze flux creep experiments it is necessary to calculate the flux transport caused by the jump rate ν_c for macroscopic sample geometries. This problem is analogous to solving a diffusion equation where the diffusion constant is a function of the potential gradient. Such a treatment was given by BEASLEY et al. (1969). They found that the total flux Φ in a cylindrical sample in a longitudinal field at the time t is given by

$$\Phi(t) = \Phi(t_o) \pm \frac{\pi}{3} k_B T R^3 \left(\frac{\partial U_{eff}}{\partial |\nabla B|}\right)^{-1} (1 \pm \delta) \, \ell n \left(\frac{t}{t_o}\right) \tag{6.4}$$

where t_o is an arbitrary reference time, R is the sample radius and ($\partial U_{eff}/\partial |\nabla B|$) is the change of the activation energy with flux density gradient (or driving force $P_D \simeq |\nabla B| B/\mu_o$). The $\pm$ signs refer to positive and negative ∇B values, corresponding to increasing and decreasing fields in the critical state. The correction factor δ takes into account the fact that $|\nabla B|$ and ($\partial U_{eff}/\partial |\nabla B|$) differ as a function of position depending upon whether the specimen is subject to increasing or decreasing fields. Usually δ is small compared to unity. The logarithmic time dependence predicted by Eq. (6.4) is in excellent agreement with the experiments (KIM et al. 1962, DUNLAP et al. 1963, BEASLEY et al. 1969, ANTESBERGER and ULLMAIER 1974, see Fig. 48).

In the approximation (6.3) for $(\partial U_{eff}/\partial|\nabla B|)$ we have for $P_D \simeq P_V$

$$\frac{\partial U_{eff}}{\partial |\nabla B|} = \frac{V_B X B}{\mu_o} = \frac{U_o B}{\mu_o P_v} \qquad (6.5)$$

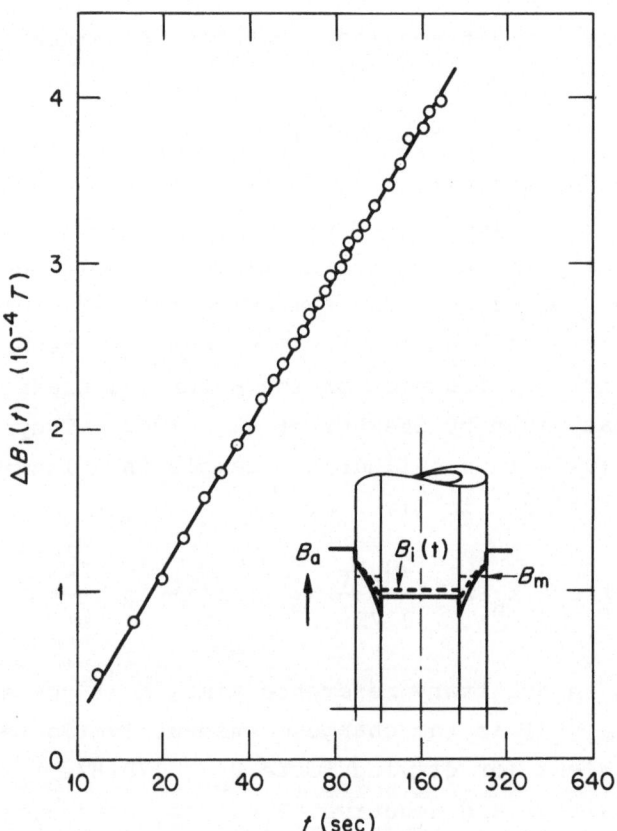

Fig.48 Time-dependence of the field inside a hollow NbTa cylinder subject to an external field which was increased from zero until it reached a value H_o at the time t = 0. Although H_o was held constant for t > 0, flux creep towards regions of smaller flux density caused the inside field B_i to increase logarithmically. Since we are only interested in field changes, the origin of the ordinate is arbitrary. The sample was identical with sample 18 in Fig.32 (ANTESBERGER and ULLMAIER 1974).

The creep rate R_c which is defined as

$$R_c \equiv \frac{d\phi(t)}{d \ln t} \tag{6.6}$$

is then given by

$$R_c = \frac{\pi}{3} k_B TR^3 \frac{\mu_o}{B \, V_B \, X} (1 \pm \delta) = \frac{\pi}{3} k_B TR^3 \frac{\mu_o P_V}{B \, U_o} (1 \pm \delta) \tag{6.7}$$

For the tube geometry of Fig. 48 a slightly different relation is obtained (see ANTESBERGER and ULLMAIER 1974). Eq. (6.7) is the central result of the analysis of Beasley, Labusch, and Webb and shows that measurements of the creep rate R_c, combined with pinning force density measurements permit the determination of the material-sensitive parameters U_o and $V_B X$ (P_V measurements alone only provide the ratio $U_o/V_B X$). Examples of the field dependence of U_o and $V_B X$ are given in Fig. 49

Unfortunately, useful experimental data on flux creep are rather scarce. Besides the pioneer work of KIM and his coworkers, only two other investigations have been reported: measurements on cold-worked PbTl alloys (BEASLEY et al. 1969), and on NbTa alloys containing normal-conducting precipitates (ANTESBERGER and ULLMAIER 1974, see Section 5.1.1). Although these experiments unambiguously confirm the existence of flux creep and are in accord with Anderson's theory, the information is not sufficient to draw detailed conclusions about the complex processes occurring during thermally activated motion of vortices. Therefore only the general characteristics of the above results are listed here: (a) The flux density gradient in a cylindrical specimen initially cycled to the critical state decrease logarithmically with time. Typical creep rates $dB_i/d \ln t$ (see Fig. 48) are around several 10^{-4} T per sec-decade for samples with 2 mm wall thickness. In a time interval from 10 to 10^4 sec (3 decades) this field change corresponds to a decrease $\Delta J_c/J_c \simeq 10^{-3}$ for $J_c = 3.10^8$ Am^{-2}. This small decay confirms the earlier statement that the concept of a stationary critical state is appropriate for almost all practical purposes. (b) Eq. (6.7) enables one to determine the barrier height U_o

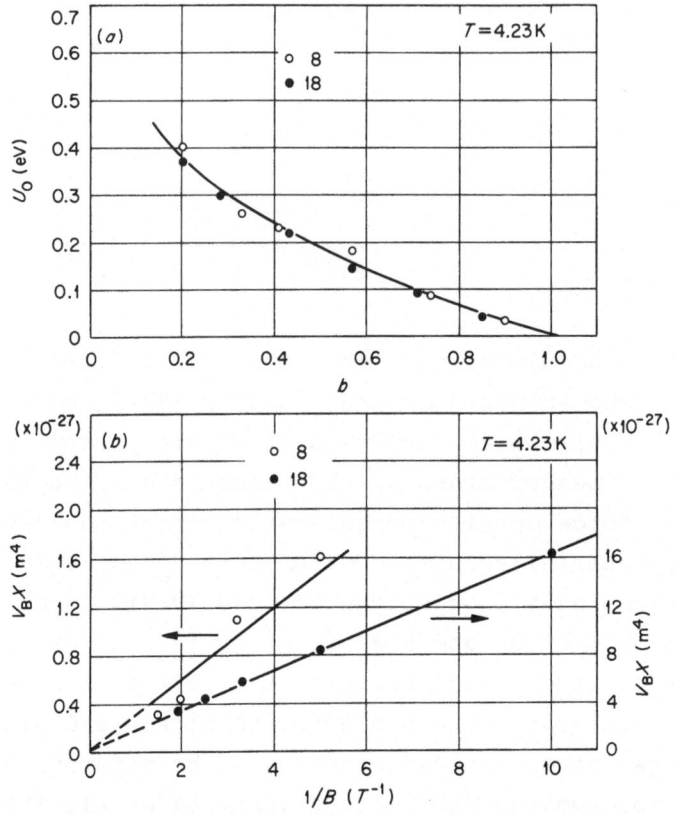

Fig.49 (a) Barrier height U_O for thermally activated motion of flux bundles as a function of the reduced flux density b for two Nb-Ta samples containing different densities of pinning centers. (b) V_BX vs the reciprocal flux density. The U_O and V_BX values where obtained from inserting experimental P_V and R_C values into Eq. (6.7) (the P_V values are shown in Fig.32).

and the product of activation volume V_B and barrier width X. Typical values for these material-sensitive parameters are $U_O \approx 0.1 - 1$ eV and $V_BX \approx 10^{-27} - 10^{-25}$ m^{-4} at intermediate fields and temperatures. As B → B_{c2}, U_O → 0, however, V_BX approaches a nonzero value (Fig. 49). (c) On changing the applied field back away from the critical state, the creep rates becomes zero and then reverses sign. (e) Flux crosses the specimen surface in increments containing from a few vortices close to H_{c2} to 100 or more at lower fields (BEASLEY et al. 1969). This "bundle" size is in agreement with noise mea-

surements in the flux flow state (VAN GURP 1968, HEIDEN and
ROCKLIN 1968). (f) There is no detactable creep in the Meissner state (H < H_{c1}).

6.2 Flux Flow and A. C. Losses

If the current through a type II superconductor in the
mixed state exceeds the critical current (case $I_2 > I_c$ in
Fig. 47), a voltage $V = E_f \cdot l$ appears along the sample, where
l is the distance between potential contacts. KIM et al. (1963,
1965) considered the electric field E_f as arising from the motion of flux lines. For several years this interpretation was
repeatedly questioned and numerous notes have been published
on this subject, both for and against this view. However, today there is direct experimental evidence that vortices do indeed move with an average velocity v_L given by Eq. (6.1) (GIAE-
VER 1966, SCHELTEN et al. 1975).

Numerous investigations of flux flow in different materials
show the same general characteristics: the voltage V_f always increases linearly with the difference $I - I_c$ and the slope of
the $V(I)$ curve is independent of the critical current I_c. In
order to describe this behavior, Kim used the following phenomenological interpretation. He assumed that the flux flow velocity v_L is determined by the balance of three forces: the
driving force P_D, the pinning force P_V, and a viscous friction
force $B\eta v_L / \phi_o$, i.e.

$$P_D = P_V + \frac{B}{\phi_o} \eta \, v_L \qquad (6.8)$$

where $\eta(B,T)$ is the "viscosity coefficient" of the superconductor. The electric field is then given from Eq. (6.1) as

$$E_f = \frac{\phi_o}{\eta} (P_D - P_V) = \frac{\phi_o B}{\eta} (J - J_c) \qquad (6.9)$$

which agrees with the experimentally observed flux flow characteristic.

Eq. (6.9) is similar in form to Ohm's law which suggests the introduction of a "flux flow resistivity" $\rho_F < \rho_n$ defined by

$$\rho_F \equiv \frac{dE_f}{dJ} = \frac{\phi_0 \, B}{\eta} \qquad (6.10)$$

A typical example for the field and temperature dependence of ρ_F is shown in Fig. 50. Of course, the phenomenological Eq. (6.9) cannot provide information about the damping mechanism since v_L was determined from Eq. (6.8) through η which was left as a free parameter. In order to clarify the physical reasons for a dissipative motion of flux lines, various damping mechanisms have been proposed. BARDEEN and STEPHEN (1965) and VAN VIJFEIJKEN and NIESSEN (1965) treated this problem theoretically on the basis of a generalized London model. For large vortex spacings (small fields) they find

$$\rho_F = \rho_n \frac{B}{B_{c2}} \qquad (6.11)$$

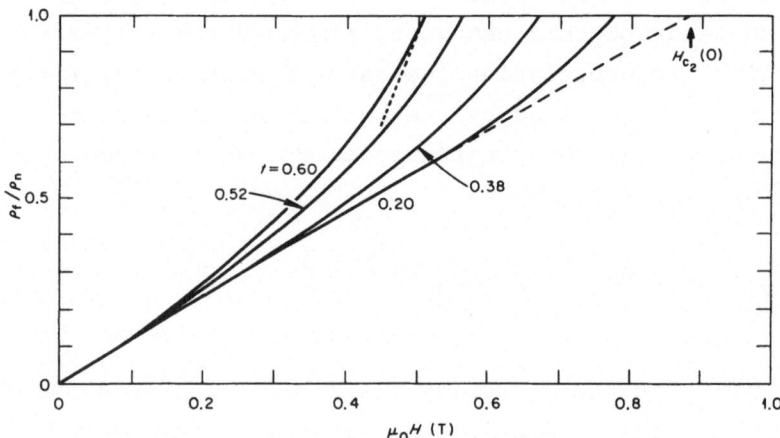

Fig. 50 Flux flow resistivity ρ_F normalized to the resistivity ρ_n in the normal state as a function of the transverse applied field H for different values of the reduced temperature t. The dashed and the dotted lines represent the theoretical predictions of Eqs. (6.11) and (6.12), respectively. $H_{c2}(0)$ is the upper critical field of the $Nb_{0.73}Ta_{0.23}$ sample at t = 0 (after BESSON et al. 1973).

106

which agrees with the experimental results, at least at low temperatures (see Fig. 50). In deriving Eq. (6.11) only losses due to the joule heat of the normal electrons have been considered. According to TINKHAM (1964), an additional dissipative mechanism caused by the finite relaxation time of the super-electrons should also be taken into account. Such relaxation processes are included in the calculation by SCHMID (1966) who applied a time dependent Ginzburg-Landau equation to the problem of vortex motion and was able to calculate ρ_F close to T_c and H_{c2}. For dirty materials ($\ell < \xi$, see Chapter 8) he finds

$$\frac{d\rho_F}{dB} = \frac{2.5 \, \rho_n}{B_{c2}(T)} \qquad (B \to B_{c2}) \qquad (6.12)$$

which agrees well with the experimental data if the temperature is not too low ($T > 0.6 \, T_c$, see Fig. 50). CAROLI and MAKI (1967, 1969) have extended Schmid's theory to both pure and dirty superconductors at arbitrary temperatures close to H_{c2}. For a more detailed review of the microscopic origin of dissipation processes during flux line motion see KIM and STEPHEN (1969) and HUEBENER (1974).

In accordance with the experiments, all the above calculations show that the viscosity coefficient η depends on parameters of the ideal type II superconductors only (i.e., on properties of the "matrix" like ρ_n, B_{c2}, etc.) and is not influenced by pinning effects (apart from the onset of flux flow which is determined by J_c). Nevertheless, flux flow effects are important for hard superconductors since they are the main reason for the energy losses observed during changes of external fields or currents. The power dissipation w per unit volume which is equal to the work done by the driving force is given by

$$w = P_D \bar{v}_L = E_f J = P_v \bar{v}_L + \frac{B}{\phi_o} \eta \, \bar{v}_L^2 \qquad (6.13)$$

The first term on the right side of Eq. (6.13) is called pinning loss because it arises from the presence of pinning centers which introduce local deformation in the vortex lattice (see Section 2.2.1). During a motion of flux lines the deformation energy tranfers to vibration energy of the flux lattice which is finally dissipated in the viscous medium. The pinning loss term was first discussed by IRIE and YAMAFUJI (1967) and is the basis for the calculation of P_v by the so-called dynamic approach (see Section 4.1). In this "microscopic" picture, the pinning loss term $P_v \bar{v}_L$ is given by (B/ϕ_o) $\eta \left[\bar{v}_L^2 - (\bar{v}_L)^2 \right]$, i.e., the dissipation is caused by the viscous damping of the local velocity fluctuations around the mean velocity $\bar{v}_L$. If there were no pinning centers, all vortices would move with a constant velocity $\bar{v}$ and it is the dissipation resulting from this motion which is given by the second term of Eq. (6.13).

At first sight it seems that measurements of the A.C. components of the flux flow voltage would yield detailed information about the velocity fluctuations caused by the pinning centers. However, theoretical considerations by CLEM (1971) and JARVIS and PARK (1971) show that the noise voltage depends on the configuration of the voltage leads and that it is necessary to assume a specific model for the flux motion before any relevant information can be extracted. Noise measurements have been performed by VAN OOIJEN (1965), VAN GURP (1968, 1969), HEIDEN and ROCKLIN (1968) and THOMPSON and JOINER (1974). A detailed discussion of noise measurements can be found in the review by HEIDEN (1971).

For technical frequencies it can be shown (ULLMAIER 1966, CAMPBELL and EVETTS 1972, p. 290) that in most cases the second term Eq. (6.13) is very small compared to $P_v \bar{v}_L$. Since the pinning loss term does not contain the viscosity coefficient explicitily, the implication is that the A.C. losses in most hard superconductors can be calculated if only the pinning force density P_v as a function of B is known. The dissipated power is obtained by integrating the power density $w \simeq P_v \bar{v}_L$ over the sample volume with $\bar{v}_L$ being expressed as a geometry-dependent function of the external field or current change. This is equivalent to the determination of hysteresis losses in magnetic materials by the area $\oint H_o dM$ under the magnetization curve. Indeed, magnetization measurements are used rather often for the determination of the pinning loss per field cycle in hard superconductors. Loss calculations for different specimen geometries which also include surface effects have been thoroughly discussed by HANCOX (1966) and WIPF (1968). Here I shall give only two results for the one-dimensional case of a plane slab in a longitudinal A.C. field which is superimposed on a static field $H_o > H_{c1}$. If the amplitude h_o of the A.C. field is small, the maximum depth $h_o/2J_c$ to which the field changes penetrate into the sample (see Fig. 7) will also be small compared to the slab thickness L. In this case the dissipated power W is given by

$$W = \frac{4\mu_o}{3} \frac{V H_o}{L P_v} h_o^3 \nu \qquad (h_o \ll J_c L) \qquad (6.14)$$

where ν is the frequency of the A.C. field and V is the specimen volume. If on the other hand h_o is so large, that the field changes fully penetrate the sample volume the loss power is

$$W = VL \frac{P_v}{H_o} h_o \nu \qquad (h_o > J_c L) \qquad (6.15)$$

In the derivation of both Eqs. (6.14) and (6.15) it was assumed that P_v was independent of B. Although the above relations hold only for the one-dimensional case, the functional dependence on P_v and h_o remains unchanged for more complicated geometries. Therefore, for small A.C. amplitudes the losses are lower when the pinning becomes stronger, however, if h_o exceeds $J_c L$ the losses begin to rise rapidly and they are greater when the pinning becomes stronger. Equation (6.15) further shows that the density of the dissipated power decreases with decreasing sample thickness and filamentary composites (see Section 5.2) would be ideal for such A.C. applications. However, unless the matrix is insulating the outer filaments effectively shield the current from the inner core. Twisting of the filaments and the use of high-resistivity matrix materials (cupro-nickel) can improve the situation (for a detailed discussion on A.C. losses in filamentary composites see WILSON et al. 1970).

In general, losses in superconductors appear to be smaller than in normal metals subject to the same field or current changes. In order to get a feeling about their order of magnitude we insert $\mu_o H_o = 1T$, $P_v = 10^9 Nm^{-3}$ ($J_c = 10^9$ Am^{-2}), $\nu = 50$ Hz, and $\mu_o h_o = 10^{-2}T$ (these are typical numbers for P_v measurements by A.C. techniques, see Section 7.3) into Eq. (6.14) and obtain $WL/V \simeq 10^{-2}$ Wm^{-2}. This is a rather small power per unit area of the sample surface which can easily be transferred into the surrounding liquid helium (the maximum heat transfer in the nucleate boiling region where the temperature difference between metal and liquid is small, is around 5×10^3 Wm^{-2} for He at 4.2 K, see Fig. 55 b).

6.3 Flux Jumps and Instabilities

With the first generation of superconducting coils it was discovered that the maximum currents which the wire in the coil windings could carry, were much lower than those predicted from the short sample characteristic (see Section 5.2) of the same wire. It was soon realized that this "degradation" phenomenon was closely connected to so-called flux jump instabilities which are latterly a consequence of the non-equilibrium nature of the critical state. Flux jumps are sudden breakdowns of the super-conducting state and have first been observed in so-called tube magnetization measurements (AUTLER 1960, KIM et al. 1962, and others, see Fig. 51). The development of such an instability can be visualized by the diagram in Fig. 52. Because hard super-

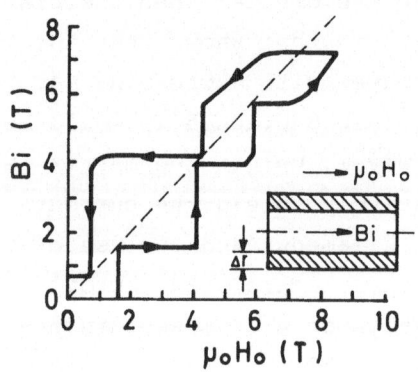

Fig.51 Magnetic field B_i inside a Nb_3Sn tube as a function of a slowly increasing and decreasing external field H_O. At certain fields the flux density gradient in the tube wall (see Fig.48) collapses and B_i becomes equal to $\mu_o H_O$ ("flux jump") (after KIM et al. 1963).

conductors are usually poor thermal conductors, any small perturbation, either internal (flux creep) or external (field or
current change, mechanical shock) which causes a power dissipation ΔW_1 gives rise to a temperature increase ΔT_1. Since the
pinning force $P_v = BJ_c$ usually decreases with increasing temperature, this results in a flux motion necessary to establish
the new critical state (broken line in Fig. 52). The flux mo-
tion in turn generates more heat ΔW_2 and a temperature rise ΔT_2.
If $\Delta T_2 > \Delta T_1$, the process will continue until the external field
fully penetrates the sample and J_c becomes zero (flux jump). If,
on the other hand, $\Delta T_2 < \Delta T_1$, the process will decay into a stable superconducting state. It is rather simple to obtain an approximate theoretical estimate of the ratio $\Delta T_2/\Delta T_1$ if one
keeps in mind that the magnetic diffusivity $D_m = \rho/\mu_o$ in alloy
superconductors is $10^3 - 10^4$ times larger than the thermal diffusivity $D_{th} = k_{th}/S_{th}$ (ρ electrical resistivity, k_{th} thermal
conductivity, S_{th} heat capacity), i.e., magnetic flux moves
much more rapidly than heat. Then, for the one-dimensional geometry of Fig. 52, the adiabatic stability criterion is obtained
as follows (see e.g., WILSON et al. 1970).

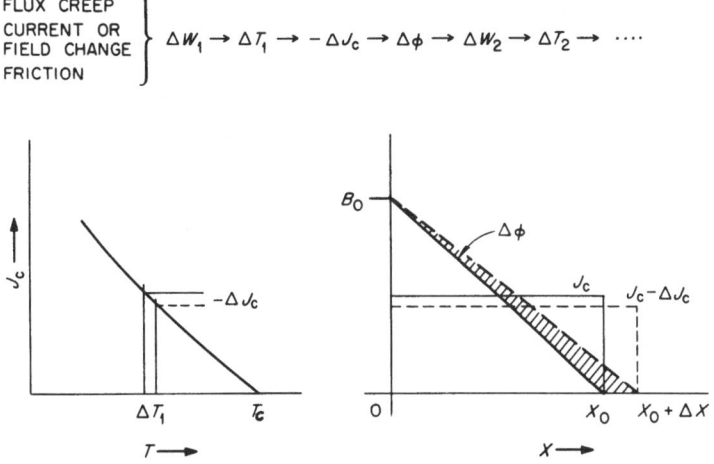

Fig.52 Growth of a small perturbation into a flux jump; distribution of flux
and current density before (full lines) and after (broken lines) a small
temperature increase in a superconducting slab (schematic).

$$\Delta J_c = \frac{dJ_c}{dT} \Delta T \tag{6.16}$$

$$\Delta x = -\frac{B_o}{\mu_o J_c} \frac{\Delta J_c}{J_c} \tag{6.17}$$

$$\Delta W_2 \text{ (average)} \simeq \frac{1}{x_o} \int_0^{x_o} \Delta \Phi \ J_c \ dx = B_o \ J_c \ \frac{\Delta x}{3} \tag{6.18}$$

$$\Delta T_2 = \frac{\Delta W_2}{S_{th}} = \frac{B_o^2}{3\mu_o \ J_c \ S_{th}} \left| \frac{dJ_c}{dT} \right| \Delta T_1 \tag{6.19}$$

Therefore the specimen will be adiabatically stable ($\Delta T_2/\Delta T_1 < 1$) if

$$B_o < (3\mu_o \ J_c \ S_{th})^{1/2} \left| \frac{dJ_c}{dT} \right|^{-1/2} \tag{6.20}$$

Since $B_o = \mu_o \ J_c \ x_o$, Eq. (6.20) is equivalent to

$$x_o < \left(\frac{3S_{th}}{\mu_o \ J_c |dJ_c/dT|} \right)^{1/2} \tag{6.21}$$

which suggests that to avoid flux jumping one should make the conductor thickness L smaller than x_o (for Nb-Ti with $S_{th} \approx 2 \cdot 10^3$ Wsec $m^{-3}K^{-1}$, $dJ_c/dT \simeq 10^9$ Am^{-2} K^{-1}, and $J_c = 3 \cdot 10^9$ Am^{-2}, Eq. (6.21) yields L < 40 μ). Although the adiabatic stabilization principle was known as long ago as 1964, there was little interest in this approach because the problems of manufacturing such fine filaments on a commercial basis seemed unsurmountable at that time.

The first technological application of adiabatic stability was achieved a few years later (CHESTER 1967a,b, SMITH 1967) with the production of the first composite conductor consisting of many filaments of NbTi in a copper matrix (for typical dimensions see Section 5.2). The high conductivity matrix assists in

removing heat from the superconductor and slows down the propagation of magnetic disturbances. In order to reduce the coupling
between filaments via the normal conducting matrix, most modern
materials have twisted filaments. Extensive experimental tests
have shown that coils wound of filamentary conductors follow the
theoretically predicted performance remarkably well (Fig. 53),
especially if the windings are vacuum impregnated in order to
prevent frictional heating from wire movement (SMITH et al. 1970).

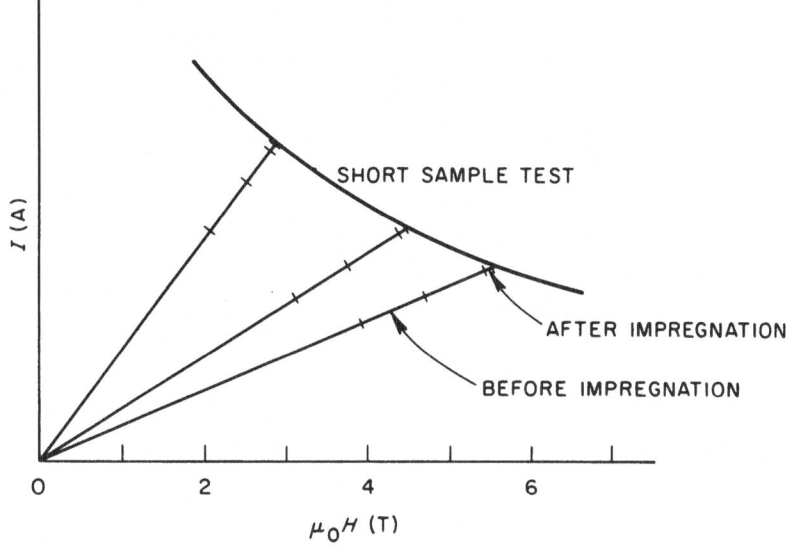

Fig.53 Typical performance of coils wound from Nb-Ti/Cu multifilament
composite lines (after SMITH et al. 1970).

Thus far adiabatic stability has not been achieved in commercial Nb_3Sn and V_3Ga tapes (for very recent progress in manufacturing multifilament wires of these materials, see the re-
marks in Section 5.2). At present, high field magnets must therefore rely on so-called dynamic stability which takes advantage
of the slowing down of the flux propagation by high conductivity
normal metal thus allowing time for most of the heat to escape
by thermal conduction (in high purity metals D_{th} is 10^3 - 10^4
times larger than D_m). A corresponding stability criterion can
most simply be derived by considering a composite formed from
alternate layers of superconductor in close contact with normal

113

metal layers of resistivity ρ_n. If the superconducting sheets have a thickness L perpendicular to the field and occupy a fraction λ_s of the total composite volume, the external field is shielded by a mean current density $\lambda_s J_c$ and extends to a depth $x_o = B_o/\mu_o \lambda_s J_c$ into the material. Proceeding as above we have

$$\Delta x = - \frac{B_o}{\mu_o \lambda_s J_c} \frac{\Delta J_c}{J_c} \tag{6.22}$$

$$\Delta T_2 \simeq \frac{dW_2}{dt} \frac{L^2}{12k_s} \simeq \frac{\Delta W_2}{\frac{1}{2}\tau_m} \frac{L^2}{12k_s} \tag{6.23}$$

where $\tau_m \simeq 4 x_o^2 \mu_o (1 - \lambda_s)/\pi\rho_n$ is the time constant for flux penetration and k_s is the thermal conductivity of the superconductor. With Eqs. (6.16) and (6.18) we then obtain for the maximum stable superconductor thickness (CHESTER 1967b)

$$L < \frac{3}{\pi} (\frac{2k_s}{\rho_n} \frac{1-\lambda_s}{\lambda_s} \frac{1}{J_c|dJ_c/dT|})^{1/2} \tag{6.24}$$

Although the dynamic stability criterion (6.24) depends on parameters different from the adiabatic criterion (6.21), it yields similar values for the maximum stable thickness of Nb-Ti (in Cu-matrix). Expressions similar to (6.24) were also derived for other conductor geometries (e.g., for an edge-cooled Nb_3Sn tape in transverse field), including the heat transfer from the tape surface to the liquid helium (HART 1968).

Both adiabatic and dynamic stabilization were concerned with the conditions under which the initiation of flux jumps is prevented. A different approach was developed by HANCOX (1966) who showed that it is possible to achieve stable coil performance without entirely eliminating flux jumps. Taking into account the

114

strong increase of the heat capacity S_{th} with temperature, an already initiated flux jump can come to a halt if at some temperate $T_2 < T_c$ the generated heat w per unit volume becomes smaller than $\int_{T_1}^{T_2} S_{th}(T)\,dT$. Under certain conditions such a partial flux jump can take place in the magnet without quenching the transport current. We can therefore distinguish between three regions in the field dependence of the critical current of a superconducting coil winding (Fig. 54): (1) stable with no flux jumps, (2) stable with partial flux jumps, and (3) unstable. Detailed calculations of the boundary curves between these regions have been carried out for different conditions. For a discussion of the results and a comparison with experimental tests the reader is referred to the review by WILSON and WALTERS (1970).

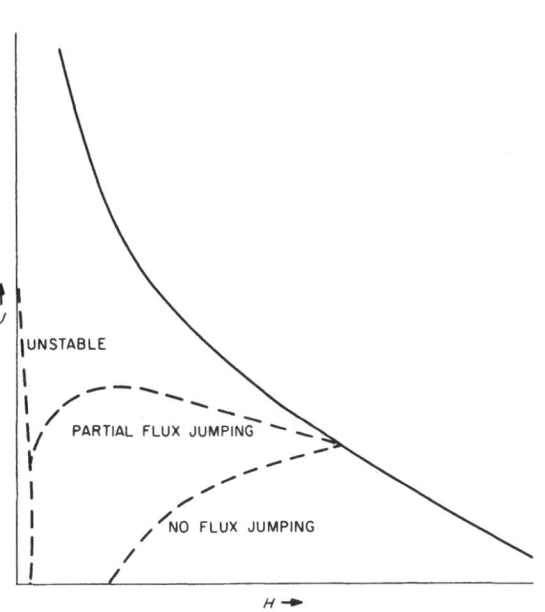

Fig.54 General form of the Hancox stability diagram showing the regions of no flux jumping, partial flux jumping and instability. The boundaries between these regions depend on the conductor thickness L and shift to higher current densities when L is decreased. The full line is the short sample characteristic of the material (from WILSON et al. 1970).

UNSTABLE

PARTIAL FLUX JUMPING

NO FLUX JUMPING

J

$H \rightarrow$

The stabilization methods discussed so far have been successfully applied to small and medium sized magnets resulting in high overall current densities (several 10^8 Am^{-2} at 5T) in the coil windings. For very large coils where the winding occupies a smaller percentage of the system volume, high overall current densities are less important. Because of the enormous stored

energies (e.g., 800 MWsec in the CERN bubble chamber magnet),
the prime requirement in large systems is insensitivity to any
form of perturbation or maloperation. This is most reliably
achieved by the so-called cryogenic stabilization (LAVERICK 1965,
KANTROWITZ and STEKLY 1965). Cryogenically stabilized coils can
be regarded as liquid helium-cooled normal state devices with
a superconducting shunt to reduce the losses to zero in steady-
state operation. Here, whatever flux jumps occur, the worst that
can happen is a local transfer of the transport current from
the superconductor to the normal substrate (Fig. 55a). If the
substrate is able to carry this current without increasing the
overall conductor temperature above a certain limit $T_1 < T_c$,
the current will begin to return into the superconductor after
the flux jump and finally the lossless current flow will be re-
stored.

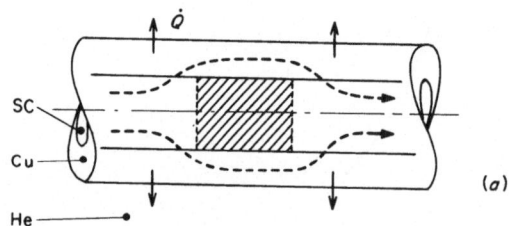

(a)

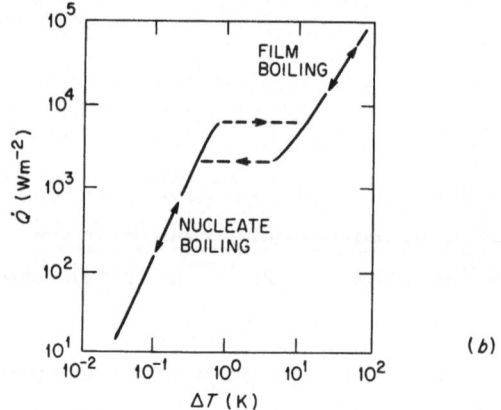

(b)

Fig.55 (a) Transfer of current to the normal conducting
substrate around a normal conducting zone in a superconducting wire. (b) Heat transfer characteristic for a ver-
tical polished metal surface
of temperature $T_b + \Delta T$ in a
helium bath of temperature T_b
(mean values from different
measurements).

116

A condition for cryogenic stability can be derived quite simply by considering the heat balance in a short portion of a normal region in an otherwise superconducting wire which is in good contact with a normal conducting substrate (Fig. 55a). Neglecting heat losses through the ends of the zone, the power balance per unit length is given by

$$\frac{I^2 \rho_n}{A} = h_t (T_1 - T_b) P \tag{6.25}$$

That is, the generated ohmic heat is equal to the heat transferred to the liquid helium (I transport current, ρ_n substrate resistivity, A cross section and P wetted perimenter of the substrate, T_b helium bath temperature). The heat transfer coefficient h_t is a strongly nonlinear function of the temperature difference $\Delta T = T_1 - T_b$ as shown in Fig. 55b. For small heat fluxes ($< 10^4$ Wm^{-2}) nucleate boiling in observed and the corresponding ΔT values are only a few tenth of a degree. At higher values of heat flux the transfer mechanism changes to film boiling with temperature differences between metal and bath between 10 and 20 K. Since this is higher than the transition temperatures of most superconductors, the composites are usually

designed so that with all the current in the substrate the maximum heat flux $Q = h_t \Delta T$ is well within the nucleate boiling regime, i.e., typically in the range $1 - 3 \times 10^3$ Wm^{-2}. Inserting this limit into Eq. (6.25) shows that the Cu/Sc ratio of a typical composite must be rather high in order to achieve cryogenic stability. For example, for $J_c = 3 \cdot 10^9$ Am^{-2} in a superconducting wire of 0.25 mm diameter, $\rho_{Cu} = 10^{-10}$ Ωm, $\dot{Q} = 2 \cdot 10^3$ Wm^{-2} we obtain a Cu/Sc ratio of around 7 (in adiabatically stable multifilament wires this ratio is only around 1.5). Together with the space requirements for the cooling channels this leads to rather low overall current densities, of the order of only a few 10^7 Am^{-2}. However, as already mentioned above, for large coils this is not a serious deficiency and the concept of cryogenic stabilization has been used very successfully in many large magnet systems.

7. Measurement of Pinning Forces

7.1 Critical Current Measurements

From the experimental standpoint, critical current measurements in which a transport current is passed through the specimen and increased until a voltage appears, is the simplest method
of determining pinning forces. However, a straightforward evaluation of such critical current measurements in terms of the
pinning force density P_v is possible only if the flux density B,
and hence P_v, does not vary significantly across the sample
cross section. This condition is fulfilled for specimens with
small cross sections, high κ-values and external fields not too
close to H_{c1}. In this case the pinning force density as a function of a transverse applied field H_0 is simply given by

$$P_v = B \, J_c \simeq B \, \frac{I_c}{A} \qquad (7.1)$$

where $B \simeq \mu_0 H_0$ is the approximately uniform flux density in the
sample of cross section A and I_c is the critical current (see
Fig. 56). The critical current is usually determined by first
applying a certain external transverse field and then increasing
the transport current through the wire or foil sample until a
certain small voltage (typically around 1 μV) appears between
two potential leads spaced a short distance apart (path OABC
in Fig. 56a). This is repeated for other H_0 values until the
whole $I_c(H)$ curve of the material is obtained ("short sample
characteristic," see also Section 5.2). If no instabilities occur, the same $I_c(H)$ curve is obtained by first applying the
transport current and then raising the field (path ODC) although

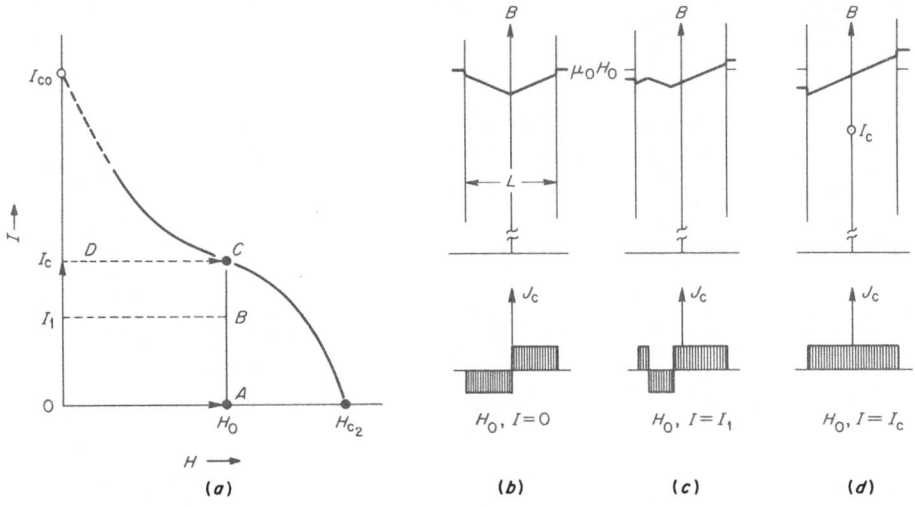

Fig.56 Flux density B and current density J_c distribution in a current carrying wire or tape in a transverse external field H_0 which is large compared to the field $I_cL/2$ produced by the transport current.

the intermediate flux distributions will then be different from those shown in Fig. 56b-d. It is interesting to note that at $I = I_c$ the irreversible part of the magnetization $B - \mu_0H$ of the sample should vanish (Fig. 56d) which has indeed been observed experimentally (LE BLANC 1963). It should also be mentioned that it is the transport current density $J_c = dH/dx$ which is measured in critical current experiments rather than the total current density J (see Eq. (2.19)). Thus, the factor dH/dB which must be taken into account in methods where dB/dx is determined (see following Sections 7.2 to 7.4), does not appear in Eq. (7.1).

At low external fields where H_0 becomes comparable to J_cL, where L is the sample diameter, it is not possible to determine P_v from critical current measurements. In this region (dotted curve in Fig. 56a), Eq. (7.1) is not applicable since the distortions of the external field due to the self-field of the current can no longer be neglected and the relation between I_c and $P_v(B)$ must be derived from the general Eq. (2.26). Finding the correct boundary conditions and integrating this differential equation is an extremely difficult problem and even for

119

simple functions $P_v(B)$ it is impossible to compute the flux distribution analytically. An exception is the special case of a current carrying wire in zero external field (KOPPE 1966). If we introduce cylindrical coordinates r, ϕ, z, it is obvious that $\underline{H}$ has only a ϕ-component which depends on r only. Then Eq. (2.26) yields

$$\frac{1}{r}\frac{d}{dr}(rH) = \frac{dH}{dr} + \frac{H}{r} = \frac{P_v(B)}{B} = J_c(B) \tag{7.2}$$

The left-hand side of this equation consists of two parts: The term dH/dr comes from the flux density gradient as in the one-dimensional case, the additional term H/r is caused by the line energy which tends to shrink a circular flux line like an extended rubber ring. Equation (7.2) shows that there is a critical radius r_c where the contracting force is just balanced by the pinning force. As soon as one compresses a circular flux line to a radius smaller than r_c it will start to contract by itself and the situation becomes unstable even if $dH/dr = 0$ at this point. We therefore have a core of radius $r < r_c$ into which no flux penetrates as long as the current is below the critical current I_{co} (Fig. 57). Since $B = 0$ and $H = H_{c1}$ at $r = r_c$, the critical radius is given by

$$r_c = H_{c1}/J_c(B = 0) \tag{7.3}$$

The critical current I_{co} can now be calculated by integrating Eq. (7.2) with the initial condition $H = H_{c1}$ at $r = r_c$ and setting

$$I_{co} = 2\pi R H(R) \tag{7.4}$$

Usually the integration must be carried out numerically. For some special functions $J_c(H)$ however analytical expressions can be given (see KOPPE 1966). The above considerations apply only for wires of radius $R > r_c$. If $R < r_c$, we can get no partial flux penetration at all and the wire behaves like a type I superconductor with a critical field H_{c1}.

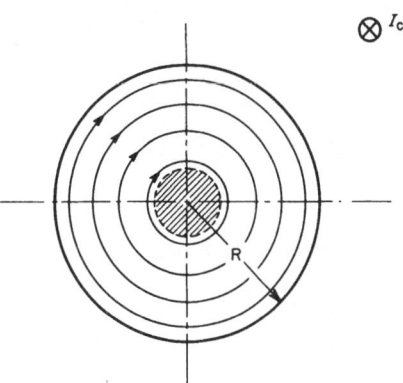

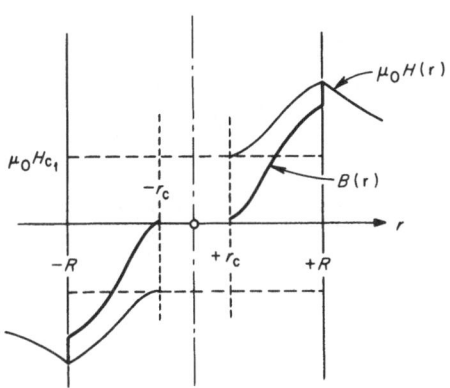

Fig.57 Magnetic field and flux density distribution in a wire in zero external field which carries the maximum transport current I_{co}. The magnetic flux in this case has penetrated the cross section exactly to the critical radius $r_c = H_{c1}/J_c(B=0)$. Circular flux lines with a radius smaller than r_c (hatched area) are unstable and will annihilate at $r = 0$.

The above theory can be tested by calculating J_{co} from Eqs. (7.2), (7.3), and (7.4) using $J_c(H)$ values obtained from another method and compare these predicted values with critical current measurements in zero-external field. This was done by ULLMAIER and KERNOHAN (1966) who found that J_{co} values predicted from magnetization measurements in longitudinal fields agreed with those measured directly in the same NbZr wires in zero field. Similar good agreement was reported by CAMPBELL et al. (1968) for PbBi samples.

Another geometry which has attracted considerable interest is the case of current carrying wire in a longitudinal field. SEKULA et al. (1963) found that longitudinal critical currents are much higher than those measured in transverse fields and in low fields even exceed the zero field value I_{co}. BERGERON (1963) suggested that such a result is caused by a force free configuration of the flux line system in which $\underline{J}$ is parallel to $\underline{B}$.

Since such configurations are thus far of academic interest only, I shall not discuss them further here but refer the reader
to the detailed discussion by CAMPBELL and EVETTS (1972, pp. 250-
265).

7.2 Magnetization Measurements

Critical current measurements which are most frequently
used in the determination of pinning forces in thin samples become very unpractical for samples with large cross sections. Besides the decreasing applicability limit of Eq. (7.1), it is
difficult to produce and control the required large currents and
lead them through the cryostat (e.g., for a 1 mm diam. wire with
$J_c = 3 \cdot 10^9$ Am^{-2}, the critical current is around 2000 A). There
can also be heating problems at the normal metal-superconductor
joints. All these difficulties are avoided in magnetization experiments in longitudinal fields where the external transport
current is replaced by "induced" circular currents for which
the critical state Eq. (2.20a) with cylindrical symmetry can be
applied. It is therefore not astonishing that most of the more
basic experimental studies of type II superconductivity made
use of magnetization measurements, at least in earlier years.

All magnetization techniques make use of the induced voltage V_i which appears during a flux change in a pick-up coil of
N turns surrounding the sample. Since the flux change
$N\Delta\Phi = NA_s <\Delta B>$ in a specimen of cross section A_s is equal to
the time integral of the induced voltage, we obtain for the
magnetization $<M> = <B> - \mu_o H_o$.

$$<M(H_o)> = -\frac{1}{NA_s} \int_{t(H=0)}^{t(H=H_o)} V_i(t')dt' - \mu_o H_o \qquad (7.5)$$

<M> is the magnetization averaged over the specimen cross sec-
tion A_s. For a cylindrical sample with radius R (Fig. 6) we
have

$$<M> = \frac{1}{\pi R^2} \int_0^B B(r) \, 2\pi \, r \, dr - \mu_o H_o \qquad (7.6)$$

which contains the sought quantity $P_v = B(dH/dB)(dB/dr)$ in an integrated form only. Thus, the evaluation of P_v from magnetization curves is in general quite difficult and time consuming. The usual way of extracting $P_v(B)$ from a $<M(H_o)>$ curve starts from making a more or less intelligent guess about the functional dependence of P_v on B. This function, which contains free parameters, and a knowledge of the reversible B(H) relation then allow a computation of flux density profiles B(r) at each external field H_o. Inserting these into Eq. (7.6) yields an expression for $<M(H_o)>$ in which the free parameters must be adjusted such that a good fit with the experimental magnetization curve is reached. Iterative methods for this procedure have been described by· IRIE and YAMAFUJI (1967) and DICHTEL et al. (1968).

The situation is greatly simplified if the hysteresis of the magnetization curve is small, i.e., if either P_v or R (or both) are small. In this case the relative difference between the maximum and minimum flux density in the sample is also small and we may assume that the pinning force density P_v is constant across the sample (see upper insert in Fig. 58). For a cylindrical specimen this leads to

$$P_v = \frac{3B}{2R} \left(\frac{dH}{dB}\right)_{rev} \left[<M(H_o\uparrow)> - <M(H_o\downarrow)> \right] \tag{7.7}$$

where $<M(H_o\uparrow)>$ and $<M(H_o\downarrow)>$ are the magnetization values in increasing and decreasing external fields, respectively. The slope $(dH/dB)_{rev}$ can be obtained from the relation $B_{rev}(H_o)$ which is given by (see Fig. 58)

$$B_{rev}(H_o) = \frac{1}{2} \left[<M(H_o\uparrow)> + <M(H_o\downarrow)> \right] \tag{7.8}$$

Because of the integral nature of magnetization curves there are a number of pitfalls which sometimes lead to a misinterpretation of experimental data, especially in low κ materials. The most common difficulty is that caused by surface currents which shield the bulk of the sample from external field changes that are smaller than a certain threshold value ΔH_s (ULLMAIER and

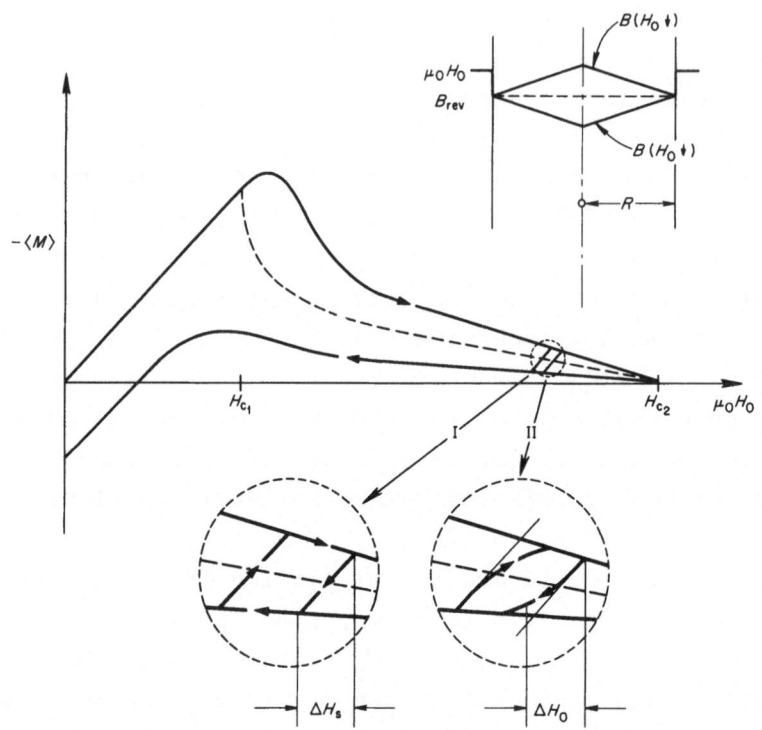

Fig.58 Irreversible magnetization curve of a type II superconductor caused by either surface currents on bulk pinning forces. Although the magnetization curves for both cases look alike, the reason for the irreversibility can be identified by the shape of minor hysteresis loops (see inserts on the bottom). The sketch on the upper right of the figure shows the flux density distribution caused by bulk pinning for an increasing $(H_O \uparrow)$ and decreasing $(H_O \downarrow)$ external field, respectively.

GAUSTER 1966). Such surface barriers can cause irreversibilities in the magnetization behavior which are quite similar to those caused by pinning effects.

It is, however, rather easy to distinguish between surface and bulk hysteresis by examining the shape of so-called minor hysteresis loops (see inserts in Fig. 58). In case I no flux changes occur in the bulk, i.e., the external field changes are shielded from the bulk by surface currents up to ΔH_s. In case II the external field change ΔH_O causes a flux change $\Delta\Phi = k\Delta H_O^2$ which corresponds to a change in the field distributions shown in Fig. 7. Since the constant k is proportional to P_v, minor hysteresis loops can also be used for a determination of P_v even for samples with strongly irreversible magnetization curves

(FREYHARDT 1969). In principle this technique is very similar to the A.C. methods which are, however, much more sensitive and which will be discussed in the next section.

In practice several methods of measuring magnetization curves or minor hysteresis loops are in use. The most common method at present is an electronic integration of the voltage from a pick-up coil around the sample which is subject to an increasing or decreasing external field (FIETZ 1965). The experimental arrangement usually contains a second coil which is necessary to cancel the signal created in the unavoidable gap between the sample and the pick-up coil. Previously, most magnetization curves were measured point by point by moving the specimen in a homogeneous stationary field from one coil into another and integrating the voltage pulse by a ballistic galvanometer. Other experimenters used magnetometers with vibrating coils or samples (FONER 1959, HEATON and ROSE-INNES 1963, GAUSTER et al. 1964). Extremely small magnetization changes (as low as 10^{-11}T in a field of 0.2 T) can be detected by incorporating a superconducting quantum interference device (SQUID). For instance, such a system was successfully used in flux creep experiments (see Section 6.1) by BEASLEY et al. (1969).

7.3 A. C. Techniques

As was first recognized by BEAN (1964), the most powerful method used in determining pinning forces is that of imposing a small ripple field on a large longitudinal D.C. field and observing the voltage V_i induced in a pick-up coil around the specimen. If one keeps the A.C. field amplitude h_o and thus the penetration depth of the flux changes small, then one may assume a constant dB/dr in the affected region near the surface of the sample (Fig. 7). This considerably simplifies the evaluation of the measurements. Small amplitudes and low frequencies are also desirable in order to keep viscous forces and heating effects negligibly small (see Section 6.2). The induced voltage signal can be processed in a number of ways, the most common being harmonic analysis (BEAN 1964, KROEGER et al. 1973), direct waveform

analysis (ULLMAIER 1966) and integration of the complete waveform using a wide-band lock-in amplifier (CAMPBELL 1969). A comparison between these three approaches shows that the last technique is by far the most versatile and is capable of revealing
details of the pinning process which cannot be obtained from any
other macroscopic pinning force measurements. For this reason
only Campbell's technique will be described here in more detail.

The flux changes in a cylindrical specimen which is subject
to a sinusoidal A.C. field (amplitude h_o) superimposed on a D.C.
field H_o are shown in Fig. 7b-e. The principle of Campbell's method
is to determine the difference in flux in the specimen at the ex-
tremes of the ripple field (triangular area in Fig. 7d). This is
achieved by feeding the voltage induced in a pickup coil around
the sample into a wide-band lock-in amplifier (phase sensitive
detector). When the phase is correctly set, its output S is proportional to the integral of the input signal over a half-cycle
of the reference waveform.

$$S = K \int_o^{T/2} V_i \, dt = kN \left[\Phi(h_o) - \Phi(-h_o) \right] \qquad (7.9)$$

N is the number of turns of the pickup coil and k is a constant
which depends on the sensitivity of the apparaturs. The flux
changes in the gap between the pick-up coil and the specimen are
cancelled by a compensating coil whose output is adjusted such
that the sum of the voltages induced in both coils is zero for
$H_o < H_{c1}$ (Meissner state). For $H_o > H_{c1}$, V_i then corresponds
to flux changes in the superconductor only. If the amplitude h_o
of the ripple field is increased by a small amount dh_o, the
field B inside the specimen will increase by dB at all points to
which the critical state has penetrated. For $h_o << H_o$, the difference between $\mu_o H_o$ and $B(R)$ at the sample surface $(r = R)$ will
be about constant within $H_o + h_o$ and $H_o - h_o$, i.e.

$$dB = \mu_o \, dh_o \qquad (7.10)$$

126

Therefore, if the critical state has reached a radius r, the increase dϕ in flux is

$$d\phi = \pi (R^2 - r^2) \mu_o \, dh_o \qquad\qquad (7.11)$$

$$\frac{d\phi}{dh_o} = \frac{1}{kN} \frac{dS}{dh_o} = \pi (R^2 - r^2) \mu_o \qquad\qquad (7.12)$$

Hence, by differentiating the amplifier output S with respect to the field amplitude h_o we can obtain the distance to which a field $\mu_o h_o$ penetrates and thus a graph showing the flux density profile as a function of distance (R - r) from the surface (Fig. 59). The critical current density at any point in the specimen may then be calculated from the slope of this curve. This procedure is valid even if J_c varies as a function of r (or B(r)). The only assumptions made in the derivation of Eq. (7.12) were (a) that Eq. (7.10) is valid, (b) that the passage of flux in and out the surface is symmetric, i.e., the diamagnetic and paramagnetic currents are equal and (c) that the flux penetration is radially symmetric.

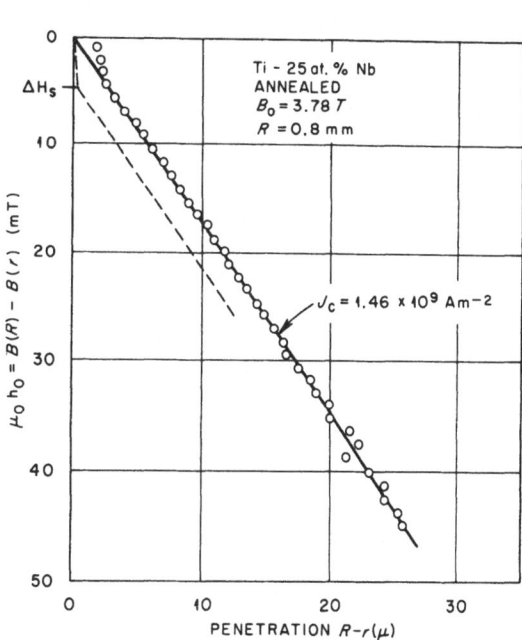

Fig.59 Flux density profile in a Ti-Nb sample in an external field of 3.78 T as determined by an A.C. technique developed by CAMPBELL (1969) (after KROEGER et al. 1975). The deviation of the experimental data from the straight solid line at small field amplitudes is due to an elastic displacement of flux which occurs before the full critical state is reached. Another possible deviation (dashed line) can be caused by a surface barrier ΔH_s.

127

Besides the pinning force density, the response to a low
amplitude A.C. field can yield information about the restoring
force on flux lines as a function of their displacement. CAMP-
BELL (1971) was able to show that the often observed initial linear and reversible response of pinned flux to a ripple field
can be explained by an elastic movement of flux lines. He found
that the restoring force F(u) as a function of the displacement
u can be approximated by

$F(u) = 2 P_V \left[1 - \exp(-u/2d_0) \right]$, i.e., the force is first linear
in u and then asymptotically approaches the maximum pinning
force which corresponds to the full critical state. For Pb-Bi
alloys containing Bi particles as pinning centers (see Section
5.1.1) Campbell measured an "interaction distance" d_0 of around
30 Å which was only weakly dependent on B. One consequence of
the displacement threshold to which vortices must be moved until they become unpinned is an initially linear increase in flux
with A.C. amplitude instead of the quadriatic increase which indeed occurs at larger amplitudes as predicted by the critical
state model (see Fig. 7). This effect can be seen in the flux
profile of Fig. 59 which is essentially the derivative of the
function $\Phi(h_0)$: at $h_0 \simeq 0$ the slope of the curve is finite and
only gradually approaches the solid line corresponding to the
critical state model.

Although the initial restoring force and the interaction
distance are not fully understood in terms of microscopic pinning parameters (U, K_0, d, etc.) there are indications that
the quantities F(u) and d_0 are closely connected with the Labusch parameter α_L which is the mean value of $\nabla\nabla U$ (U being the
local interaction energy per unit length of a flux line, see
Section 2.2). This interesting relation which would open the
way to a measurement of α_L is discussed in detail by CAMPBELL
(1971) and CAMPBELL and EVETTS (1972, p. 360).

Other deviations from the critical state model where the
flux distribution in the whole sample is determined by just one
parameter $P_V(B)$, are caused by surface effects of various kinds
(see for example dashed line in Fig. 59). A.C. measurements are
in general an ideal tool to examine such effects. There are,
however, cases for which the response of the superconductor to

128

an A.C. field cannot yet be satisfactorily explained (e.g., samples with polished surfaces and low bulk pinning forces)..

Very recently a systematic comparison of the results from four methods of measuring pinning forces has been made by KROE-GER et al. (1975). Measurements by each method were carried out on several samples, having a range of κ values, critical current densities, and diameters. The techniques used for determining J_c were: (a) critical current measurements, (b) magnetization, (c) harmonic analysis, and (d) total flux A.C. measurements. In general good agreement between the different techniques is obtained (Fig. 60) but one must beware of minor failures of the critical state hypothesis. The authors therefore suggest that the best results are likely to be obtained by employing more than one method.

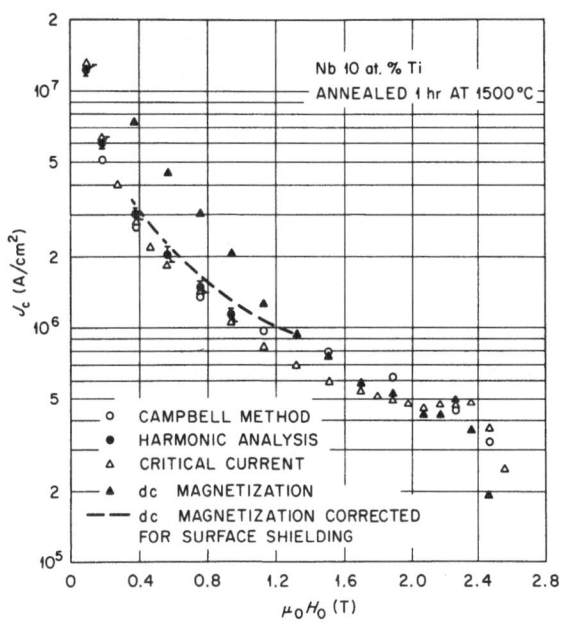

Fig.60 Comparison of critical current densities J_c measured by different techniques in the same sample (after KROEGER et al. 1975).

7.4 Other Techniques

One of the first techniques used for investigations of pinning forces in hard superconductors was that of measuring the magnetic fields inside and outside a tube sample (KIM et al. 1962, see Fig. 51). If the wall thickness Δr of the tube is sufficiently small one may assume that the flux density gradient is constant within the material. For this case the pinning force density is simply given by

$$P_v = B J_c \simeq \pm B_m \frac{H_o - H_i}{\Delta r} \qquad (7.13)$$

where B_m is the mean flux density, H_i is the magnetic field in the hole of the tubular sample and the $\pm$ signs correspond to increasing or decreasing external fields H_o, respectively. H_i is usually determined by a small Hall probe. Besides its simplicity, the method has the advantage that it also permits a convenient study of flux creep phenomena (Section 6.1) and instabilities (Section 6.3 and Fig. 51).

A small flat Hall probe can also be used to measure the flux profile $B(r)$ directly by moving it in a thin transverse slot inside the superconductor (GAUSTER et al. 1965). Although this method demonstrates the critical state most directly, it is difficult to obtain accurate quantitative results. This is mainly a result of the damage introduced in the surrounding material during the machining of the slot. Another uncertainty which comes from the field distortions caused by the slot can be corrected to a certain extent (KOPPE 1965b). Direct measurements of the flux profile have also been reported by COFFEY (1967), VOIGT (1968), SIKORA (1968), and WEBER and RIEGLER (1973).

An experimental estimate of pinning forces can also be obtained from a direct measurement of the force needed to unpin the flux lines. Such mechanical measurements which usually employ a disc-shaped sample as part of a torsion pendulum were described by HEISE (1964), WIPF (1964), WRAIGHT (1971), and EGGENDORFER (1973). Although the results clearly show the validity of the concepts of vortex pinning, viscous flow forces, etc., reliable values of the pinning force are difficult to obtain because of the large field distortions necessary to unpin the flux lines. Furthermore, it is impossible to distinguish between surface and bulk pinning effects.

All methods discussed so far yield more or less accurate information about the bulk pinning force density P_v. First steps toward an investigation of individual pinning processes are a direct examination of the vortex distribution by the decoration technique (ESSMANN and TRÄUBLE 1969) and neutron depolarization experiments (WEBER et al. 1971 und WEBER 1974). However, both of these methods encounter severe difficulties in the quantitative evaluation of the results. The decoration technique suffers from the presence of a free surface and from forces due to the line tension of bent vortices comparable to those from the flux gradient, which are the only ones that can be determined by the technique. In the case of polarized neutron transmission the theoretical grounds for relations connecting the measured degree of depolarization to the distortions of the vortex lattice by the pinning centers are not yet well established. At present it is therefore difficult to draw any quantitative conclusions from these interesting results. A third way of investigating individual pinning interactions is by means of neutron diffraction. Since this method is free of most of the disadvantages mentioned above, it could possibly be developed into a powerful tool for the study of pinning processes. The principle of the technique and preliminary results are discussed in the last section.

7.5 Neutron Diffraction

As the result of a proposal by DE GENNES and MATRICON (1964), the first neutron diffraction experiments in type II superconductors were performed by CRIBIER et al. (1974). These experiments provided the first direct proof of the existence of a regular array of flux lines, each line carrying a single flux quantum. Important technical improvements in the last few years made it possible to apply the neutron small angle diffraction technique to study (1) the microscopic field distribution in the mixed state (SCHELTEN et al. 1971, 1972), (2) the morphology of flux lattices at rest and in motion (SCHELTEN et al. 1974 and 1975, THOREL et al. 1973), and (3) the mutual misalignment of flux lines (LIPPMANN et al. 1973).

The neutron diffraction method is based on the interaction of the magnetic moment of the neutrons with the spatially vary-

ing magnetic field of the mixed state structure. With a periodic arrangement of the scattering centers (i.e., the flux lines)
the scattered neutrons interfere in such a way that intensity
maxima appear only at certain angles θ_{hk} given by the Bragg
equation

$$2d_{hk} \sin \theta_{hk}/2 = \lambda_n \qquad (7.14)$$

Taking typical values for the flux lattice spacings $d_{hk} \simeq 1000$ Å
and the neutron wavelength $\lambda_n \simeq 10$ Å, diffraction angles of
around 30' are obtained for (hk) = (10) (hk denotes the index
of the lattice plane). Such small angles require large distances ($\sim$ 10 m) between sample and detector plane in order to se-
parate reflections of different order sufficiently.

In principle a diffraction pattern can be obtained from two
different geometries. With the neutron beam perpendicular to the
flux lines, properties of the flux lattice in planes perpendicular to the field direction can be studied. With the beam almost parallel to the vortices, information about their mutual
misalignment can be obtained from an analysis of so-called
rocking curves (Fig. 61). A rocking curve is obtained by recording the scattered intensity I_θ at a fixed scattering angle θ
as a function of the rocking angle ϕ.

It is evident that the width of such a rocking curve is
closely connected to the misalignment angles of the flux lines
with respect to the external field direction, which are in turn
related to the distortions u_0 caused by the individual pinning
forces K. Neutron diffraction experiments can therefore provide
information about the individual interactions between flux
lines and pinning centers. From macroscopic pinning force measurements (Sections 7.1-7.4) this information can be extracted
only through theoretical relations which are still afflicted
with uncertainties and limitations.

There is also an essential difference in the experimental
situation between neutron diffraction and macroscopic pinning
force measurements. In the latter case, external field or cur-

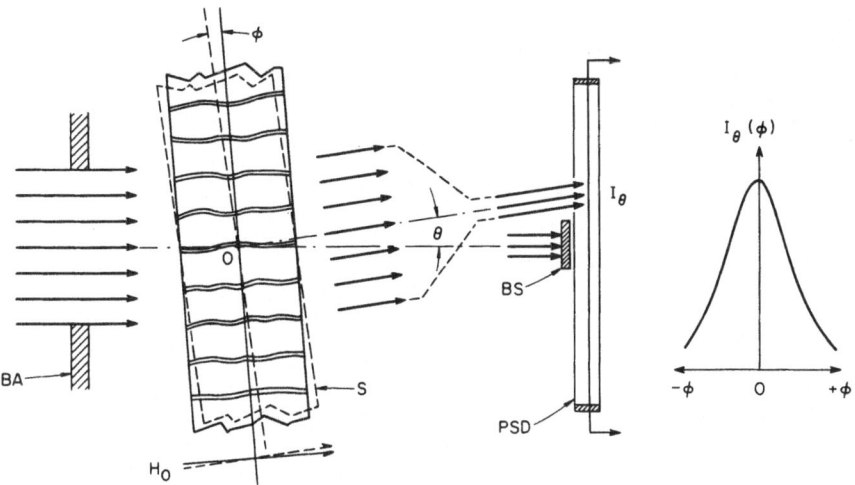

Fig.61 Experimental set up (schematic top view) for the measurement of the mutual misalignment of flux lines by neutron small-angle diffraction. For the sake of clarity, the sample region is drawn on an enlarged scale. In reality the distance between sample S and position sensitive detector PSD ($\sim$10 m) is about 10^4 times larger than the sample thickness ($\sim$1 mm). The direct beam is blocked by the beam stop BS. On the right the dependence of the diffracted intensity I_θ on the rotation angle ϕ is shown. The width of this rocking curve is a measure of the misalignment angles of the flux lines.

rent changes create macroscopic flux density gradients which are related to the pinning force density P_v through Eq. (2.20). In such a sample the flux lattice spacings, d_{hk} vary over a wide range and no well defined Bragg reflections can be observed. This obstacle can be overcome by cooling the sample in a constant magnetic field B/μ_o from a temperature above T_c to the required temperature $T < T_c$. During this process flux lines are created close to their final positions and no macroscopic flux density gradients are generated, i.e. the flux density is the same throughout the sample (apart from a thin surface layer of thickness $\simeq$ $(B-B_{rev})/\mu_o J_c)$.

Flux lattices grown by this procedure show sharp diffraction lines at angles θ_{hk} given by Eq. (7.14) with $d_{hk} = (\sqrt{3}\ \phi_o/2B)^{1/2}$. The only action of the pinning centers in the material is now the generation of local distortions of the vortex arrangement without supporting a macroscopic driving force. Preliminary calculations by LABUSCH (1974) show that these distortions lead to

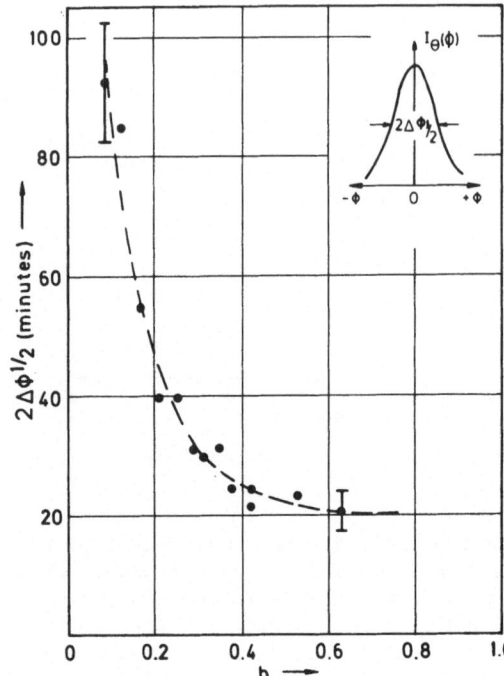

Fig.62 The full width 2 $\Delta\phi_{1/2}$ at half maximum of rocking curves as a function of the reduced flux density b for a Nb sample containing 1.7 x 10^{18} Nb_2N precipitates per m^3 (B_{c2} = 0.461 T, κ = 1.63) at T = 4.2 K (after LIPPMANN 1974).

a full width 2 $\Delta\phi_{1/2}$ at half maximum of the rocking curve (see Fig. 61 and insert in Fig. 62) given by

$$2\Delta\phi_{1/2} = \frac{\tau_L}{8\pi} N_v'(B) <K^2> \frac{1}{C_{44} C_{66}}$$ (7.15)

Here τ_L = $2\pi/d_{hk}$ is the length of the reciprocal lattice vector of the vortex lattice, $N_v'(B)$ is the density of pinning active centers and K is the interaction force between a flux line and a pinning center. C_{44} and C_{66} are the elastic constants of the flux lattice as defined in Section 2.2.1.

Very recently some first experimental results have been reported (LIPPMANN 1974). Since the quantitative evaluation of this data is still in a preliminary stage, I present here only a typical field dependence of 2 $\Delta\phi_{1/2}$. The curve in Fig. 62 was measured in a Nb single crystal containing normal conducting Nb_2N precipitates, i.e. the pinning structure was similar to that of the macroscopic measurements in NbTa alloys described in Section 5.1.1 (Figs. 32-34). Taking into account the different H_{c2} and κ values of the two materials one finds good agreement between the K values in Fig. 33 and the values obtained by inserting measured $2 \Delta\phi_{1/2}$ values into Eq. (7.15).

The full width at half maximum is of course only a small part of the information contained in a rocking curve and it is clear that a detailed analysis of the results will be based on the entire intensity distribution $I_\theta(\phi)$. It is planned to apply the neutron diffraction technique to a series of samples covering a wide concentration range of well-characterized pinning centers. In connection with simultaneous macroscopic P_v measurements, an evaluation of the results by means of theoretical expressions relating $I_\theta(\phi)$ to microscopic pinning parameters (like α_L, d, u_o, etc.) should lead to an improved understanding of flux pinning in type II superconductors.

8. Appendices

A. A Brief Excursion into Equilibrium Properties of the Mixed State

In thermodynamic equilibrium the Gibbs free energy G_s of a superconductor at constant temperature and external field is a minimum. The difference between the Gibbs free energy G_n in the normalconducting state and G_s in the superconducting state is determined by two terms: the first is due to the condensation of a small part of the conduction electrons into Cooper pairs and leads to a decrease in energy; the second is caused by the expulsion of magnetic flux from the sample volume and leads to an increase in G. If we disregard surface effects, these two terms cancel each other at a certain external magnetic field (the thermodynamic critical field H_c), i.e. the Gibbs free energies of both phases become equal and the superconductor undergoes a transition into the normal state.

In the presence of phase boundaries this consideration must be modified. Figure A.1 shows schematically the behavior of the density n_s of Cooper pairs and the magnetic field h in the boundary region between a normalconducting and a superconducting zone. Changes of n_s and h cannot occur abruptly but will extend over certain distances. The characteristic lengths are the coherence length ξ and the penetration depth λ for variations of n_s and h, respectively.

Figure A.1a is valid for the case $\xi \gg \lambda$. Here the magnetic field in the hatched volume is almost completely expelled, i.e. the positive contribution to G is close to its maximum value. On the other hand, the density of Cooper pairs is still rather low and there will be no significant reduction of G, i.e. the sum

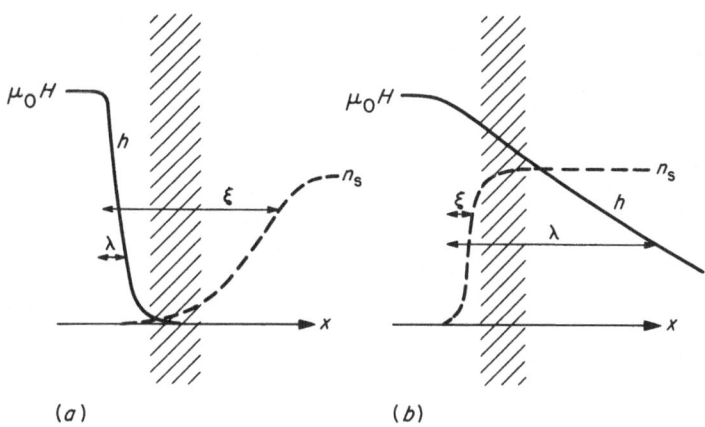

Fig.A1 Density of Cooper pairs n_s and magnetic field h in the boundary region between a normal-conducting and a superconducting zone for (a) a type-I and (b) a type-II material (λ penetration depth, ξ coherence length).

of the two contributions from the boundary will be positive and the Gibbs free energy in this region will be larger than G_s in the bulk of the superconductor. Thus materials with $\xi \gg \lambda$ have a positive surface energy and are called type I superconductors. They will reach a minimum in G if the area of boundaries between normal and superconducting regions is as small as possible. The smallest area is the physical boundary of the specimen, therefore the bulk of a type I material will be in a uniform (super-conducting) state where the external magnetic field $H_o < H_c$ is expelled (MEISSNER state; see Fig. A.2a).

The situation is quite different for the case $\xi \ll \lambda$ which is schematically shown in Fig. A.1b. Here the energy conserving condensation term involving n_s has already reached its full value in the hatched region, whereas the positive flux expulsion term involving h is still small. Therefore the Gibbs free energy in the boundary will be smaller than G_s, i.e. superconductors with $\xi \ll \lambda$ have a negative surface energy and are called type II su-perconductors. If a magnetic field is applied to such a material, it can lower its free energy by breaking up into many alterna-ting superconducting and normalconducting zones. This state is called SHUBNIKOV phase or mixed state. Macroscopically it shows

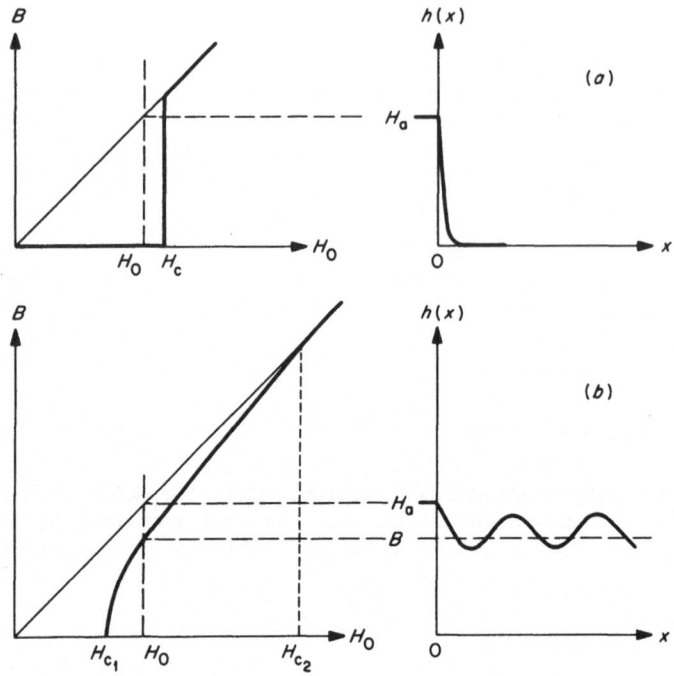

Fig.A2 Equilibrium flux density B as a function of the external field H_o in (a) a type-I and (b) a type-II superconductor. B is the average value of the microscopic field h which is shown on the right hand side of the figure as a function of the distance x in the superconductor.

up in a partial flux penetration, the average flux density B being at some value between zero (MEISSNER state) and H_o (normal state; see Fig. A.2b).

The above considerations provide a simple qualitative explanation of the existence of two kinds of superconductors. However they give no indication about the geometrical shape of the coexisting normal and superconducting zones in the mixed state. In principle there are two ways to maximize the surface to volume ratio: lamina of small thickness or filaments of small diameter ($\geq \xi$). Theory shows that the filament structure is lowest in energy and is indeed always found experimentally. The occurrence of such a structure was first predicted by ABRIKOSOV (1952, 1957). He also showed that whether a material is type I or type II superconducting depends on the so-called Ginzburg-

Landau parameter κ which is defined as the ratio of the two characteristic lengths λ and ξ. Type I superconductivity is found for $\kappa = \lambda/\xi < 1/\sqrt{2}$, and type II behavior is observed for $\kappa > 1/\sqrt{2}$.

Each filament of a mixed state structure consists of a nor- malconducting core of radius $\approx \xi$ and, together with its super- conducting surrounding (radius $\approx \lambda$) is called a flux line or a vortex. An example of the structure of a flux line is given in Fig. A.3 where the density of Cooper pairs n_s, the magnetic field h, and the vortex current density j_s as a function of the distance from the line center (r = 0) is shown.

A flux line is equivalent to a superconducting ring and it is known (F. LONDON 1950) that the magnetic flux contained in such a multiconnected system is quantized in units of

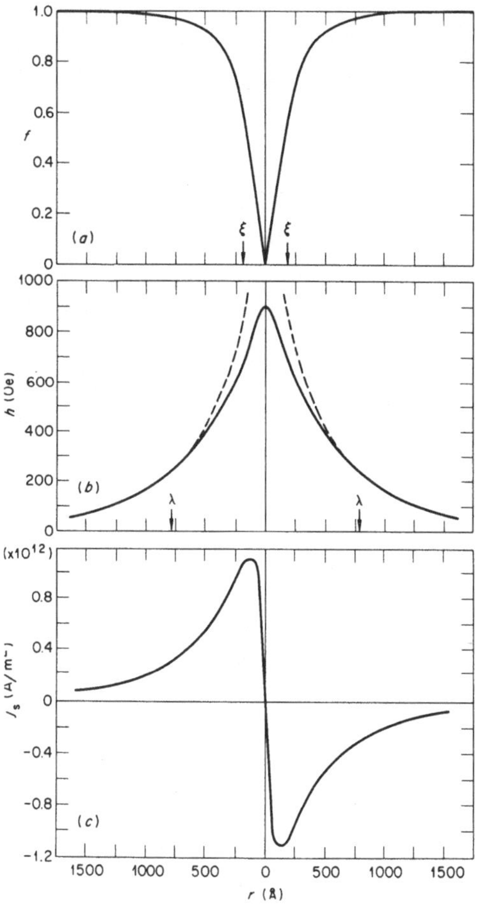

Fig.A3 Normalized order parameter $f = \psi/\psi_o$, magnetic field h, and current density j_s as a function of the distance r from the center of an isolated flux line in $Nb_{73}Ta_{27}$ at 4.2 K. h(r) and j_s(r) = dh/dr were determined from neu- tron diffraction measurements (SCHELTEN et al. 1971). Apart from the core region, the London model (dotted lines in (b)) provides a good description of the structure of a flux line in this material.

$\phi_o = \dfrac{h}{2e} = 2.07 \times 10^{-15}$ Vsec.

Since the energy per unit length of a flux line depends quadratically on the flux ϕ contained in it (see Eq. 8.11), a configuration of vortices each carrying the smallest possible amount of flux, i.e. a single flux quantum ϕ_o, is energetically most favorable and it is therefore always observed that

$$\phi = \int_o^\infty h(r)\ 2\ \pi r dr = \phi_o \tag{8.1}$$

This leads to a simple relation between the macroscopic flux density B, the density of vortices n, and the distance a_o between lines in a triangular lattice

$$B = n\phi_o = \dfrac{2\ \phi_o}{\sqrt{3}\ a_o^2} \tag{8.2}$$

(e.g. for B = 1000 G, a_o = 1530 $\mathring{A}$).

In describing the mixed state in a quantitative way it is practical to divide the field range between H_{c1} and H_{c2} into two regions: at low fields nearer to H_{c1} the mixed state structure is rather accurately described by the so-called London-model, in the high field range nearer to H_{c2} the Ginzburg-Landau theory is more applicable. In the following some results of both approaches are listed:

(1) The LONDON-model is valid to a good approximation at all temperatures and electron mean free paths, but is restricted to materials with high Ginzburg-Landau parameters ($\kappa \gg 1$) and to the field range $H_{c1} < H \ll H_{c2}$. The microscopic magnetic field $\underline{h}$ is described by

$$\underline{h} + \lambda^2\ \text{curl curl}\ \underline{h} = \underline{\phi}_o\ \delta_2(\underline{r}), \tag{8.3}$$

where $\underline{\phi}_o$ is a vector in the direction of $\underline{h}$ with magnitude ϕ_o and $\delta_2(\underline{r})$ is a two-dimensional δ-function. In Eq. (8.3) the flux line core is reduced to a line discontinuity ($\xi = 0$; see Fig. A.4a) and we are left with only one characteristic length λ, the penetration depth, which is given by

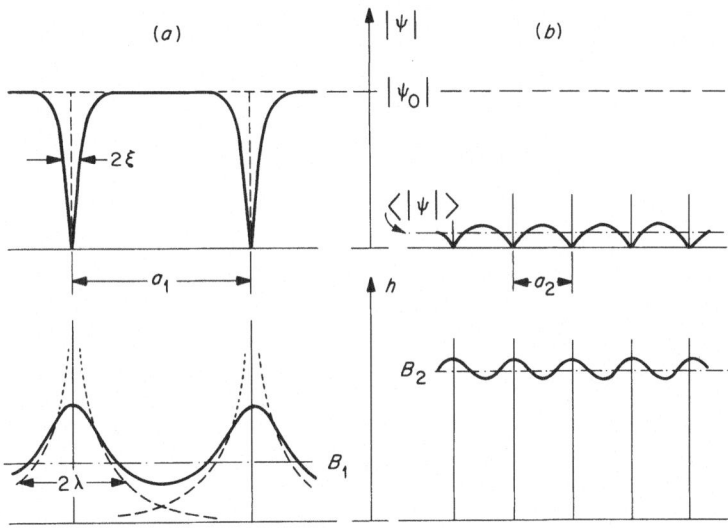

Fig.A4 Spatial distribution of the order parameter ψ and the magnetic field h in the mixed state (schematic). At low flux densities B, (case (a):$a_1 \gg \xi$), the mixed state structure can be obtained from a linear superposition of the fields of individual vortices. At high flux densities B_2 (case (b):$a_2 \simeq \xi$), the concept of individual vortices breaks down because ψ nowhere reaches its maximum value ψ_0.

$$\lambda = (m/\mu_0 \, e^2 \, n_s)^{1/2} \, .$$ (8.4)

Together with div $\underline{h}$ = 0, Eq. (8.3) yields for the field of an isolated vortex line

$$h(r) = \frac{\phi_0}{2\pi \, \lambda^2} \, K_0 \, (\frac{r}{\lambda})$$ (8.5)

and for the current density

$$j_s(r) = \frac{\phi_0}{2\pi \, \mu_0 \, \lambda^3} \, K_1 \, (\frac{r}{\lambda}) \, .$$ (8.6)

K_0 and K_1 are the MacDonald functions of order zero and one, respectively. Field distributions determined by neutron scattering experiments are in reasonable agreement with Eq. (8.5) for $r \gg \xi$ (see Fig. A.3). They also show that the London model is unable to describe the core region correctly which is of course expected from the use of $\delta_2(\underline{r})$ in Eq. (8.3).

141

With the asymptotic behavior for K_0 in Eq. (8.5) we have

$$h(r) = \frac{\phi_0}{2\pi \lambda^2} \ln \left(\frac{\lambda}{r}\right) \qquad \text{for } \xi < r << \lambda \qquad (8.7)$$

and

$$h(r) = \frac{\phi_0}{2\pi \lambda^2} \left(\frac{\pi \lambda}{2r}\right)^{1/2} e^{-r/\lambda} \qquad \text{for } r >> \lambda \qquad (8.8)$$

The energy E_L per unit length of an isolated flux line of length L is the sum of the magnetic field energy and the kinetic energy of the vortex currents

$$E_L = \frac{1}{L} \int \left(\frac{1}{2\mu_0} h^2 + \frac{1}{2} mn_s v^2\right) dV \qquad (8.9)$$

Replacing the superelectron velocity $\underline{v}$ by the current density $\mu_0 \underline{j}_s = \text{curl } \underline{h}$, Eq. (8.9) can be written as

$$E_L = \frac{1}{2\mu_0 L} \int (h^2 + \lambda^2 |\text{curl } \underline{h}|^2) dV \qquad (8.10)$$

Integration of this equation and the use of Eq. (8.7) for $h(\xi)$ finally leads to a line energy per unit length of

$$E_L = \frac{1}{4\pi \mu_0} \left(\frac{\phi_0}{\lambda}\right)^2 \left(\ln \frac{\lambda}{\xi} + \varepsilon\right) \qquad (8.11)$$

In this equation the term ε has been added in order to account for the condensation energy required to produce the normal-conducting flux line core. This contribution to E_L which is usually small ($\varepsilon \sim 0.1$) was neglected in eq. (8.9).

The Gibbs free energy density g of a system of n vortices per unit volume is

$$g = nE_L + \sum_{ij} U_{ij} - (BH - \mu_0 H^2) \qquad (8.12)$$

where U_{ij} describes the repulsive interaction between flux lines. For two parallel vortices at positions $\underline{r}_i$ and $\underline{r}_j$, respectively, U_{ij} is given by

$$U_{ij} = \frac{\phi_o \, h_{ij}}{\mu_o} \tag{8.13}$$

with

$$h_{ij} = h_i(\underline{r}_j) = h_j(\underline{r}_i) = \frac{\phi_o}{2\pi \, \lambda^2} K_o \left(\frac{|\underline{r}_i - \underline{r}_j|}{\lambda}\right) \tag{8.14}$$

From the expressions (8.12) and (8.13) the B(H) relation (which is equivalent to the reversible magnetization curve) is obtained with the condition $\partial g / \partial B = 0$. The results can be expressed in closed form for some limiting cases:

For very low flux densities ($a_o \gg \lambda$), the interaction term can be ignored and from

$$g = B \left(\frac{E_L}{\phi_o} - H\right) + \mu_o H^2 \quad \text{and} \quad \frac{\partial g}{\partial B} = 0 \tag{8.15}$$

we obtain the lower critical field H_{c1} where the first flux penetration occurs

$$H_{c1} = \frac{E_L}{\phi_o} = \frac{\phi_o}{4\pi \, \mu_o \, \lambda^2} \, \ell n \, \frac{\lambda}{\xi} \tag{8.16}$$

For the case $a_o > \lambda$ we may still neglect the flux line interactions except for that of the nearest neighbors and write

$$g = B \left(H_{c1} + \frac{z}{2} \frac{\phi_o}{2\pi \, \mu_o \, \lambda^2} K_o \left(\frac{a_o}{\lambda}\right) - H\right) + \mu_o H^2 \tag{8.17}$$

This leads to

$$H = H_{c1} + \frac{z}{2} \frac{\phi_o}{4\pi \, \mu_o \, \lambda^2} \left(\frac{\pi a_o}{2\lambda}\right)^{1/2} e^{-a_o/\lambda} \tag{8.18}$$

143

z is the number of nearest neighbors ($z = 6$ for a triangular lattice).

For medium flux densities $\xi \ll a_o < \lambda$ the interaction term must include contributions from rather distant neighbors and a similar calculation yields

$$H = \frac{B}{\mu_o} + H_{c1} \frac{\ln(0.23\, a_o/\xi)}{\ln(\lambda/\xi)} \qquad (8.19)$$

Equations (8.18) and (8.19) agree well with the results of magnetization measurements on materials with high Ginzburg-Landau parameters (see Fig. A.5).

The above considerations are based upon a linear superposition of the fields of individual flux lines and on the assumption of a spatially homogeneous Cooper pair density n_s (except in the line discontinuities at the vortex centers; see Fig. A.4a). It is obvious that these assumptions become inadequate at high flux densities where the distance between vortices is comparable to the core dimensions ($a_2 \geq \xi$; see Fig. A.4b).

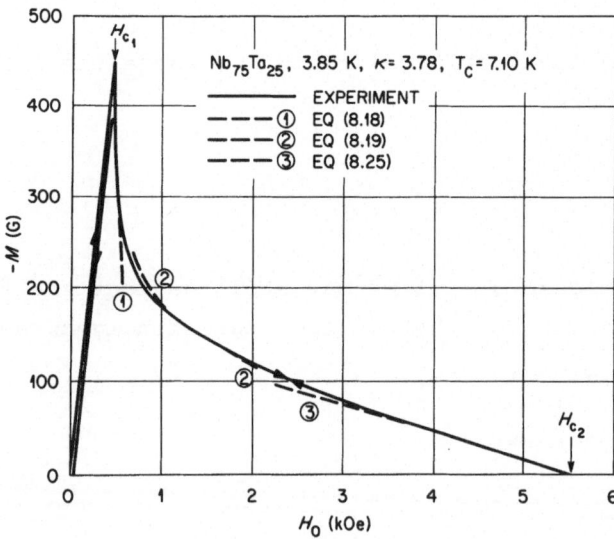

Fig.A5 Comparison of measured and calculated magnetization behavior of a NbTa alloy. The theoretical curves 1 and 2 are based on the London model, curve 3 is obtained from the Ginzburg-Landau theory (ULLMAIER and KERNOHAN 1970).

(2) In the GINZBURG-LANDAU theory the superconducting state is characterized by the so-called order parameter ψ which is related to the Cooper pair density by $|\psi|^2 = n_s$. Close to T_c or close to H_{c2} where ψ is small (Fig. A.4b), the difference between the free energy in the normal and superconducting state can be written (GINZBURG and LANDAU 1950) as

$$f_s - f_n = \alpha |\psi|^2 + \frac{\beta}{2}|\psi|^4 + \frac{1}{2m}\left|(i\hbar\nabla - \mu_o e^+ \underline{A})\psi\right|^2 + \frac{\mu_o}{2}h^2 \qquad (8.20)$$

where curl $\underline{A} = \underline{h}$, $e^+ = 2e$, and α and β are constants which can be expressed in terms of H_{c2} and κ

$$\alpha = -\left(\frac{e\hbar}{m}\right)\mu_o H_{c2}$$

$$\beta = 2\mu_o \left(\frac{e\hbar}{m}\right)^2 \kappa^2 \qquad (8.21)$$

$$|\psi_o|^2 = -\frac{\alpha}{\beta} = \frac{1}{2}\left(\frac{m}{e\hbar}\right)\frac{H_{c2}}{\kappa^2}$$

ψ_o is the order parameter in zero field. The first two terms on the right side of Eq. (8.20) are the first terms of a power series expansion in $|\psi|^2$, the third term is the kinetic energy of the charge carriers and the last term is the magnetic field energy. Minimizing $f_s - f_n$ for variations of ψ and $\underline{h}$ leads to the famous Ginzburg-Landau equations

$$\frac{1}{2m}(-i\hbar\nabla - 2e\,\mu_o\,\underline{A})^2 \psi + \alpha\psi + \beta|\psi|^2 \psi = 0$$

$$\text{curl } \underline{h} = \underline{j}_s = \frac{e\hbar}{im}(\psi^{\times}\nabla\psi - \psi\nabla\psi^{\times}) - \frac{4\mu_o e^2}{m}\psi\psi^{\times}\underline{A} \qquad (8.22)$$

The first relation describes the behavior of ψ in the precence of a magnetic field, the second gives the current density distribution. A general solution of these equations, even on a numerical basis, is a difficult problem. However, it is possible to find analytical solutions in the low field limit (London model) and in the high field limit. This was first demonstrated by

ABRIKOSOV (1952, 1957) and led to the recognition of type II superconductivity. Some important results of his calculations in the high field case $H < H_{c2}$ are:

(a) The upper critical field H_{c2} is given by

$$H_{c2} = \kappa\sqrt{2}\ H_c = \frac{\phi_o}{2\pi\ \mu_o\ \xi^2} \tag{8.23}$$

Below H_{c2} the microscopic magnetic field h is related to the order parameters ψ by

$$h = \mu_o \left[H - \frac{H_{c2}}{2\kappa^2} \left(\frac{\psi}{\psi_o}\right)^2 \right] \tag{8.24}$$

where ψ_o is the order parameter in zero magnetic field, given by Eq. (8.21).

(b) ψ and h are periodic functions of the Cartesian coordinates x and y in a plane perpendicular to the field direction z.

(c) The macroscopic flux density $B = <h>$ varies linearly with the external field $H = H_o$

$$B = \mu_o \left[H - \frac{H_{c2}-H}{(2\kappa^2-1)\ \beta_A} \right] \tag{8.25}$$

For large κ-values where $B \simeq \mu_o H$ we may combine Eqs. (8.24) and (8.25) to obtain

$$\left< \left|\frac{\psi}{\psi_o}\right|^2 \right> = 1 - \frac{B}{\mu_o H_{c2}} = 1 - b \tag{8.26}$$

i.e. the mean condensation energy which is proportional to $n_s = |\psi|^2$ decreases linearly with the reduced flux density b. Finally, the average value of the Gibbs free energy density difference is

$$<g_m(H,T) - g_n> = -\frac{\mu_o}{2}\ \frac{(H_{c2}-H)^2}{(2\kappa^2-1)\ \beta_A} \tag{8.27}$$

where the indices m and n refer to the mixed and normalconduc-
ting state, respectively. The ratio $\beta_A \equiv <|\psi|^4>/|<\psi^2>|^2$ is a
function of the geometry of the vortex arrangement only. For
a triangular lattice $\beta_A = 1.16$, for a square lattice $\beta_A = 1.18$.
This shows that a triangular flux arrangement will be lowest in
energy (see Eq. 8.27), at least in isotropic superconductors.[+]
The phenomenological Ginzburg-Landau theory was related to
the microscopic BCS theory by GORKOV (1959, 1960). This provided the link between the constants α and β in Eq. (8.20) and
the physical constants of the metal. GORKOV also showed that
the characteristic lengths λ and ξ strongly depend on the electron mean free path ℓ of the material. In Table A.1 some useful

Table A.1

	Pure Limit ($\ell \to \infty$)	Dirty Limit ($\ell \ll \xi_o$)
Scale of spatial variations of ψ	$\xi(T) = 0.74\, \xi_o \left(\dfrac{T_c}{T_c-T}\right)^{1/2}$	$\xi(T) = 0.85 \left(\dfrac{\xi_o \ell T_c}{T_c-T}\right)^{1/2}$
Scale of spatial variation of h	$\lambda(T) = \dfrac{\lambda_L(0)}{\sqrt{2}} \left(\dfrac{T_c}{T_c-T}\right)^{1/2}$	$\lambda(T) = 0.64\, \lambda_L(0) \left(\dfrac{\xi_o T_c}{\ell T_c-T}\right)^{1/2}$
Ginzburg-Landau parameter $\kappa=\lambda/\xi$	$\kappa = \kappa_o = 0.96\, \dfrac{\lambda_L(0)}{\xi_o}$	$\kappa = 0.75\, \dfrac{\lambda_L(0)}{\ell}$
Correlation length between current and vector potential	$\xi_p = \xi_o$	$\xi_p = \ell$

results of the GLAG (Ginzburg-Landau-Abrikosov-Gorkov) theory valid in the clean ($\ell \to \infty$) and dirty ($\ell \ll \xi_o$) limits are listed.

In the above relations the constants ξ_o and $\lambda_L(0)$ defined by

$$\xi_o = \frac{0.18 \ v_F}{k_B T_c} \quad \text{and} \quad \lambda_L(0) = \left(\frac{1}{2\mu_o e^2 v_F^2 N(0)}\right)^{1/2} \tag{8.28}$$

have been used. v_F is the Fermi velocity and $N(0)$ is the density of states at the Fermi surface. For the dirty limit κ can be expressed in a very useful form in terms of the electronic specific heat coefficient γ and the electrical resistivity ρ_n in the normal state (GOODMAN 1962)

$$\kappa = \kappa_o + A \ \gamma^{1/2} \ \rho_n \tag{8.29}$$

κ_o is the Ginzburg-Landau parameter of the pure material and $A = 7.5 \times 10^6$ if γ is in Wsec m^{-3} K^{-2} and ρ_n is in Ωm.

The fact that some materials are type I and others are type II superconductors can be understood from a study of Table A.1. In simple (nontransition) pure metals the Fermi velocity v_F is high and the electron mass is small ($m \approx m_o$), i.e. ξ_o is large and $\lambda_L(0)$ is small (e.g. for Al ξ_o = 16,000 Å and $\lambda_L(0)$ = 160 Å). These metals show type I behavior. On the other hand, pure transition metals (Nb, V, Tc) and intermetallic compounds of the β-tungsten type (Nb$_3$Sn, V$_3$Ga, Nb$_3$Ge, etc.) have small Fermi velocities, large effective masses, and high transition temperatures. Therefore, ξ_o can be as small as 40 Å and $\lambda_L(0)$ is large (up to 2000 Å), making these materials type II superconductors.

Alloys, in which ξ and λ are modified by mean free path effects (dirty limit in Table A.1) are usually always type II superconductors since ℓ becomes smaller than $\lambda_L(0)$ even at rather low impurity concentrations (e.g. Ta (type I in pure state) becomes type II if doped with as little as 0.15 at. % N (AUER and ULLMAIER 1973)).

As already mentioned, the GLAG theory which is based on the Ginzburg-Landau equations, is rigorous only in a restricted

domain near the transition temperature. The first important ex-
tention of the GLAG theory is the dirty limit theory ($\ell \ll \xi_0$)
of MAKI (1964) and DE GENNES (1964), valid at arbitrary temperatures near H_{c2}. These calculations show that the GLAG theory
is applicable throughout this region if the Ginzburg-Landau
parameter κ is replaced by two temperature-dependent parameters
given by

$$\kappa_1 = \frac{H_{c2}(T)}{\sqrt{2}\, H_c(T)} \tag{8.30}$$

$$\kappa_2 = \frac{1}{\sqrt{2}} \left[1 + \frac{1}{\beta_A \left(dB/\mu_0 dH - 1 \right) \Big|_{H_{c2}}} \right]^{1/2} \tag{8.31}$$

Other extensions in the region near H_{c2} for arbitrary mean free
paths and anisotropies of impurity scattering have been carried
out by HELFAND and WERTHAMER (1966), MAKI and TSUZUKI (1965)
and EILENBERGER (1967). Further progress was made possible by
Eilenberger's reformulation of the Gorkov theory in terms of
simpler transport-like equations. Using this approach, Kramer
and Pesch were able to provide a complete description of type II
behavior for all temperatures, fields, Ginzburg-Landau parameter and mean free paths. In a series of papers they present numerical calculations of flux line structures, transport coefficients, specific heat, density of states, etc. For a list of
references of this recent work, see KRAMER and PESCH (1974).

B. List of Symbols and Units

a_0	flux line spacing	m
b	reduced flux density ($b = B/B_{c2}$)	-
b_0	Burgers vector	m
d	interaction range of pinning center	m

d_o	interaction distance	m
e	electron charge	Asec
f	free energy density	Nm^{-2}
g	Gibbs free energy density	Nm^{-2}
h	local magnetic field	T
h_o	amplitude of ripple field	Am^{-1}
$h, \hbar$	Planck's constant	$Nmsec^2$
h_t	heat transfer coefficient	$Wm^{-2}K^{-1}$
j	local current density	Am^{-2}
j_s	surface current density	Am^{-2}
k	constant	-
k_B	Boltzmann constant	NmK^{-1}
k_{th}	thermal conductivity	$Wm^{-1}K^{-1}$
ℓ	spacing between pinning centers	m
m	(free) electron mass	kg
n	flux line density ($n = B/\phi_o$)	m^{-2}
n_c	number of defects in pinning "cluster"	-
n_s	density of Cooper pairs	m^{-3}
p	pressure	Nm^{-2}
r, ϕ, z	cylindrical coordinates	m
s^+	transport entropy density	$Nm^{-2}K^{-1}$
t	time	sec
u	distortion of flux lattice	m
u_o	equilibrium distortion of flux lattice	m
v_F	Fermi velocity	$msec^{-1}$
v_L	flux line velocity during flux flow	$msec^{-1}$
w	loss power density	Wm^{-3}
x, y, z	Cartesian coordinates	m

A	area	m^2
A_c	area of flux lattice unit cell	m^2
$\underline{A}$	vector potential	Am^{-2}
B	macroscopic flux density	$T=Vsec\,cm^{-2}$
<B>	flux density averaged over sample cross section	T
B_o	equilibrium flux density	T
B_{c2}	upper critical flux density $(B_{c2} = \mu_o H_{c2})$	T
C_{ij}, C'	elastic constants of the flux lattice	Nm^{-2}
C_e	effective modulus of the flux lattice	Nm^{-2}
D	microstructure dependent constant in scaling laws	$Nm^{-3}T^{-n}$
D_m	magnetic diffusivity	$m^2 sec^{-1}$
D_{th}	thermal diffusivity	$m^2 sec^{-1}$
E	electric field strength	Vm^{-1}
E_f	electric field caused by flux flow	Vm^{-1}
E_L	flux line energy per unit length	N
F	free energy	Nm
F	"macroscopic" force per unit area	Nm^{-2}
G_n	Gibbs free energy in the normal state	Nm
G_s	Gibbs free energy in the superconducting state	Nm
$G'(O)$	u_o/K for point forces	mN^{-1}
H	magnetic field strength $(H = df/dB)$	Am^{-1}
H_o	external magnetic field	Am^{-1}
H_c	thermodynamic critical field	Am^{-1}
H_{c1}	lower critical field	Am^{-1}
H_{c2}	upper critical field	Am^{-1}
ΔH_s	surface barrier field	Am^{-1}
I	current	A
I_c	critical current	A

I_{co}	critical current in zero external field	A
I_s	surface current per unit length	Am^{-1}
J	macroscopic current density	Am^{-2}
J_c	critical current density	Am^{-2}
J_m	magnetization current density	Am^{-2}
K	interaction force between flux line and pinning center	N
K_o	maximum value of K	N
L	linear dimensions of pinning centers, samples, etc.	m
M	magnetization	T
$<M>$	magnetization averaged over sample cross section	T
M_{rev}	reversible magnetization	T
M'_{eff}	effective mass of a flux line per unit length	kgm^{-1}
N	number of turns in coils	-
N_A	number of pinning centers per unit area	m^{-2}
N_v	number of pinning centers per unit volume	m^{-3}
$N(0)$	electron density of states at Fermi level	$N^{-1}m^{-4}$
P	macroscopic force density	Nm^{-3}
P_D	driving force per unit volume	Nm^{-3}
P_L	lattice force per unit volume	Nm^{-3}
P_v	pinning force per unit volume	Nm^{-3}
Q	heat flux	Wm^{-2}
R	radial distance, sample radius	m
R_c	creep rate	Vsec/decade
S	entropy	NmK^{-1}
S^+	transport entropy	NmK^{-1}
S_{44}	shear compliance	$N^{-1}m^2$

ΔS_{44}	difference of S_{44} between superconducting and normal state	$N^{-1}m^{-2}$
S_{th}	heat capacity	$Wsec\,cm^{-3}K^{-1}$
T	absolute temperature	K
T_c	superconducting transition temperature	K
T_L	flux line tension	Nm^{-1}
U	pinning potential per unit length	N
U_o	maximum value of U	N
U_{eff}	barrier height for flux creep	Nm
U_{ij}	interaction energy between flux lines per unit length	N
V	volume	m^3
V_B	activation volume for flux creep	m^3
V_i	voltage induced by flux changes	V
W	loss power	W
X	width of energy barrier for flux creep	m
α_H	Hall angle	–
α_L	Labusch parameter ($\alpha_L \equiv <\nabla\nabla U>$)	Nm^{-2}
β_A	$<\|\psi\|^4>/\|<\psi^2>\|^2$	–
γ	electronic specific heat coefficient	$Nmkg^{-1}K^{-2}$
δ	correction factor for flux creep	–
ε_j	strains in the flux lattice	–
ε_v	volume dilatation between mixed and normal state	–
ζ	misfit parameter of inclusions in the metal lattice	–
η	viscosity coefficient for flux flow	$Nsec\,cm^{-2}$
θ	scattering angle	–
κ	Ginzburg Landau parameter	–
κ_o	Ginzburg Landau parameters of pure superconductors	–

λ	penetration depth	m
λ_L	London penetration depth	m
μ_o	permeability of vacuum	$VsecA^{-1}m^{-1}$
ν	Poisson's number, A.C. frequency	- , Hz
ν_c	jump rate of flux bundles during flux creep	sec^{-1}
ξ	coherence length	m
ξ_o	correlation length in pure superconductors	m
$\rho(K)$	density of pinning interactions with force K	N^{-1}
ρ_{FD}	flux line dislocation density	m^{-2}
ρ_f	flux flow resistivity	Ωm
ρ_n	normal state resistivity	Ωm
σ_i	stresses in the flux lattice	Nm^{-2}
σ_L	stresses in the crystal lattice	Nm^{-2}
τ	relaxation time	sec
Φ	magnetic flux	Vsec
Φ_o	flux quantum	Vsec
ϕ	misalignment angle of flux lines	-
ψ	superconducting order parameter	m^{-1}
ψ_o	value of ψ in zero magnetic field	m^{-1}

9. References

The literature cited for this monograph has been collected approximately until the middle of 1974. For the most recent progress in this field I would draw the reader's attention to the papers presented at the International Discussion Meeting on Flux Pinning in Superconductors, held in September 1974 in St. Andreasberg, Germany. The proceedings of this conference are expected to appear in the spring of 1975 (Prof. P. Haasen, Institut für Metallphysik der Universität, D 34 Göttingen, Germany).

ABRIKOSOV, A.A. (1952) Dokl. Akad. Nauk SSSR 86, 489

ABRIKOSOV, A.A. (1957) Zh. Eksperim. i Teor. Fiz. 32, 1442

ALDEN, T.H. and LIVINGSTON, J.D. (1966) J. Appl. Phys. 37, 3551

ALERS, G.A. and WALDORF, D.L. (1961) Phys. Rev. Lett. 6, 677

ANDERSON, P.W. (1962) Phys. Rev. Letters 9, 309

ANTESBERGER, G. and ULLMAIER, H. (1974) Phil. Mag. 29, 1101

ANTESBERGER, G. and ULLMAIER, H. (1975) Phys. Rev. Letters 35, 59

APPLEYARD, J.R., EVETTS, J.E., and CAMPBELL, A.M. (1974) Solid State Comm. 14, 567

AUER, J. and ULLMAIER, H. (1973) Phys. Rev. B7, 136

AUTLER, S.H. (1960) Rev. Sci. Instr. 31, 369

BAKER, C. and SUTTON, J. (1969) Phil. Mag. 19, 1223

BARDEEN, J. and STEPHEN M.J. (1965) Phys. Rev. 140, A1197

BEAN, C.P. (1962) Phys. Rev. Letters $\underline{8}$, 250

BEAN, C.P. (1964) Rev. Mod. Phys. $\underline{36}$, 36

BEAN, C.P. and LIVINGSTON, J.D. (1964) Phys. Rev. Letters $\underline{12}$, 16

BEASLEY, M.R., LABUSCH, R. and WEBB, W.W. (1969) Phys. Rev. $\underline{181}$,682

BERGERON, D.J. (1963) Appl. Phys. Letters $\underline{3}$, 63

BERNDT, H., KARTASCHEFF, N. and WENZL, H. (1968) ZS. Angew. Physik $\underline{24}$, 305

BESSON, A., AOMINE, T., and RINDERER, L. (1973) J. Low Temp. Phys. $\underline{11}$, 289

BIBBY, G.W. (1970) unpublished (see CAMPBELL and EVETTS, 1972, p. 332)

BRÄNDLI, G., ENCK, F.D., FISCHER, E. and OLSEN J.L. (1968) Helv. phys. Acta $\underline{41}$, 706

BRANDT, E.H. (1969a) Phys. Stat. Sol. $\underline{36}$, 371; $\underline{36}$, 393; $\underline{36}$, K167

BUCKEL, W. (1972) Supraleitung, Physik Verlag, Weinheim

CAMPBELL, A.M. (1969) J. Phys. C $\underline{2}$, 1492; (1971) $\underline{4}$, 3186

CAMPBELL, A.M. and EVETTS, J.E. (1966) Proc. of the 10th Internat. Conf. Low Temp. Phys. Vol IIB (Moscow: Veniti) p. 26

CAMPBELL, A.M. and EVETTS, J.E. (1972) Adv. Physics $\underline{21}$, 199

CAMPBELL, A.M., EVETTS, J.E. and DEW HUGHES, D. (1968) Phil. Mag. $\underline{18}$, 313

CAROLI, C. and MAKI, K. (1967) Phys. Rev. $\underline{164}$, 591; (1969) $\underline{169}$, 381

CHANG, C.C., MCKINNON, J.B. and ROSE-INNES, A.C. (1969) Phys. Stat. Sol. $\underline{36}$, 205

CHESTER, P.F. (1967a) Proc. 1st Intern. Cryogenic Eng. Conf., Tokyo (Ed. K. Mendelssohn) Heywood-Temple Publi. p. 147

CHESTER, P.F. (1967b) Repts. Progr. Phys. $\underline{30}$, 561

156

CLEM, J.R. (1971) Physica 55, 377

COFFEY, H.T. (1967) Cryogenics 7, 73

COOTE, R.I. (1970) Thesis Cambridge (see CAMPBELL and EVETTS 1972, p. 383)

COOTE, R.I., EVETTS, J.R., and CAMPBELL, A.M. (1972) Can. J. Phys. 50, 421

CRIBIER, D., JACROT, B., MADHAW, RAO L. and FARNOUX, B. (1964) Phys. Letters 9, 106

DE GENNES, P.G. (1964) Physik Kond. Mat. 3, 79

DE GENNES, P.G. (1966) Superconductivity of Metals and Alloys, Benjamin, New York

DE GENNES, P.G. and MATRICON, J. (1964) Rev. Mod. Phys. 36, 45

DEUTSCHER, G., and DE GENNES, P.G. (1969) in: Superconductivity (Ed. R.D. Parks) M. Dekken, Inc., New York, Vol. II, p. 1005

DEW-HUGHES, D. (1974a) Phil. Mag. 30, 293

DEW-HUGHES, D. (1974b) 6th Annual Spring Meeting of AIME, Pittsburg, Pa., May 1974

DEW-HUGHES, D., and WITCOMB, M.J. (1972) Phil. Mag. 26, 73

DICHTEL, K., KOPPE, H., and SEKULA, S.T. (1968) Phys. Letters 27A, 174

DUNLAP, R.D., HEMPSTEAD, C.F., and KIM, Y.B. (1963) J. Appl. Phys. 34, 3147

EGGENDORFER, G. (1973) J. Low Temp. Phys. 10, 715

EILENBERGER, G. (1967) Phys. Rev. 153, 584

ESSMANN, U. and SCHMUCKER, R. (1974) Phys. Stat. Sol. (b) 64, 605

ESSMANN, U. and TRÄUBLE, H. (1967) Phys. Letter, A26, 526

ESSMANN, U. and TRÄUBLE, H. (1969) Phys. Stat. Sol. 32, 337

EVETTS, J.E. (1974) Intern. Discussion Meeting on Flux Pinning in Superconductors, St. Andreasberg/Harz, Germany, Sept. 23-27

EVETTS, J.E. and CAMPBELL, A.M. (1966) Proc. 10th Intern. Conf. Low Temp. Phys. Vol. IIB (Moscow: Veniti) p. 33

EVETTS, J.E., CAMPBELL, A.M., and DEW-HUGHES, D. (1968) J. Phys. C $\underline{1}$, 715

EVETTS, J.E. and WADE, J.M.A. (1970) J. Phys. Chem. Sol $\underline{31}$, 973

FARRELL, D.E., CHANDRASEKHAR, B.S., and HUANG, S. (1968) Phys. Rev. $\underline{176}$, 562

FETTER, A.L., and HOHENBERG, P.C. (1969) in: Superconductivity, Vol. II (Ed. R.D. Parks) M. Dekker Inc., New York, pp. 817-924

FETTER, A.L., HOHENBERG, P.C. and PINCUS, P. (1966) Phys. Rev. $\underline{147}$, 140

FIETZ, W.A. (1965) Rev. Sci. Instr. $\underline{36}$, 1621

FIETZ, W.A., and ROSNER, C.H. (1974) IEEE Transactions on Magnetics $\underline{10}$, No. 2

FIETZ, W.A., and WEBB, W.W. (1969) Phys. Rev. $\underline{178}$, 657

FIORY, A.T. and SERIN, B. (1966) Phys. Rev. Letter $\underline{16}$, 308; (1971) Physica $\underline{55}$, 73

FONER, S. (1959) Rev. Sci. Inst. $\underline{30}$, 548

FREYHARDT, H.C. (1969) Z. Metallk. $\underline{60}$, 409

FREYHARDT, H.C. (1971a) Phil. Mag. $\underline{23}$, 369 and (1971b) $\underline{23}$, 345

FREYHARDT, H.C. (1974) personal communication

FRIEDEL, J., DE GENNES, P.G., and MATRICON, J. (1963) Appl. Phys. Letter $\underline{2}$, 119

GAUSTER, W.F., EFFERSON, K.R., COFFEY, D.L., and SIMPKINS, J.E. (1965) ORNL Report 3908, p. 137

GAUSTER, W.F., TODD, J.H., and ULLMAIER, H. (1964) ORNL Report 3760, p. 91

GIAEVER, I. (1966) Phys. Rev. Letters $\underline{15}$, 825

GINZBURG, V.L., and LANDAU, L.D. (1950) Zh. Eksperim. i. Teor. Fiz. $\underline{20}$, 1064

GOOD, J.A., and KRAMER, E.J. (1970) Phil. Mag. $\underline{22}$, 329; (1971) $\underline{24}$, 339

GOODMAN, B.B. (1961) Phys. Rev. Letters $\underline{6}$, 597

GOODMAN, B.B. (1962) IBM J. Res. Develop. $\underline{6}$, 63

GORKOV, L.P. (1959) Zh. Eksperim. i. Teor. Fiz. $\underline{36}$, 1918

GORKOV, L.P. (1960) Soviet Phys. JETP $\underline{10}$, 593 and 998

GORTER, C.J. (1962) Phys. Letters $\underline{1}$, 69 and $\underline{2}$, 26

GREGORY, E., MARANCIK, W.G., and ORMAND, F.T. (1974) Applied Superconductivity Conference, Oakbrook, USA, Oct. 1974

HAASEN, P. (1970) Nachr. Akad. Wissensch. Göttingen, II. Mathem. Naturwissensch. Klasse, p. 10

HALLER, T.R. and BELANGER, B.C. (1971) IEEE Trans. Nucl. Sci. $\underline{18}$, 671

HAMPSHIRE, R.G., and TAYLOR, M.T. (1972) J. Phys. F $\underline{2}$, 89

HANCOX, R. (1966) Proc. IEE $\underline{113}$, 1221 and Culham Lab. Report No. CLM-P121

HANAK, J.J., and ENSTROM, R.E. (1966) Proc. of the 10th Intern. Conf. Low Temp. Phys., Vol. III (Moscow: Veniti) p. 10

HART, H.R. (1968) Proc. Brookhaven Summer Study on Superconductivity Devices, p. 571

HEARMON, R.F.S. (1961) An Introduction to Applied Anisotropic Elasticity, Oxford University Press, Oxford, p. 63

HEATON, J.W., and ROSE-INNES (1963) J. Sci. Instr. $\underline{40}$, 369

HEIDEN, C. (1971) Habilitationsschrift Universität Münster

HEIDEN, G.I., and ROCKLIN, E. (1968) Phys. Rev. Letters 21, 691

HEISE, B.H. (1964) Rev. Mod. Phys. 36, 64

HELFAND, E., and WERTHAMER, N.R. (1966) Phys. Rev. 147, 288

HILL, D.C., MORRISON, D.D., and ROSE, R.M. (1969) J. Appl.
Phys. 40, 5160

HILLMAN, H. and HAUCK (1972) Proc. Appl. Superconductivity Conf.
Annapolis, Md., May 1972, p. 429

HUEBENER, R.P. (1974) Physics Reports 13C, 143

IRIE, F., and YAMAFUJI, K. (1967) J. Phys. Soc. Jap. 23, 255

JARVIS, P., and PARK, J.G. (1971) Physica 55, 386

JOSEPHSON, B.D. (1966) Phys. Rev. A152, 211

KAMMERER, U. (1969) Phys. Stat. Sol. 34, 81

KANTROWITZ, A.R. and STEKLY, Z.J.J. (1965) Appl. Phys. Letters
6, 56

KERNOHAN, R.H. (1965) personal communication

KIM, Y.B., HEMPSTEAD, C.F., and STRNAD, A.R. (1962) Phys. Rev.
Letters 9, 306

KIM, Y.B., HEMPSTEAD, C.F., and STRNAD, A.R. (1963) Phys. Rev.
129, 528 and (1965) 131, 2486

KIM, Y.B., and STEPHEN, M.J. (1969) in: Superconductivity (Ed.
R.D. Parks) M. Dekker, Inc., New York, Vol. II, p. 1107

KOCH, C.C., and KROEGER, D.M. (1974) Phil. Mag. 31, 27

KOPPE, H. (1965a) ORNL Report 3908, p. 140

KOPPE, H. (1965b) ORNL Report 3908, p. 140

KOPPE, H. (1966) Phys. Stat. Sol. 17, K 229

KRAMER, E.J. (1970) J. Appl. Phys. 41, 621 and (1973) 44, 1360

KRAMER, E.J., and BAUER, C.L. (1967) Phil. Mag. 15, 1189

KRAMER, L., and PESCH, W. (1974) J. Low Temp. Phys. 15, 125

KROEGER, D.M., KOCH, C.C., and COGHLAN, W.A. (1973) J. Appl. Phys. $\underline{44}$, 2391

KROEGER, D.M., KOCH, C.C., and CHARLESWORTH, J.P. (1975) to be published in J. Low Temp. Phys.

KRONMÜLLER, H., and SEEGER, A. (1969) Phys. Stat. Sol. $\underline{34}$, 781

KUNZLER, J.E., BUEHLER, E., HSU, F.S.L., and WERNICK, J.H. (1961) Phys. Rev. Letters $\underline{6}$, 89

KUSAYANAGI, E., and YAMAFUJI, K. (1969) Phys. Letters $\underline{A29}$, 529

LABUSCH, R. (1966a) Phys. Letters $\underline{22}$, 9

LABUSCH, R. (1966b) personal communication, see also Ref. (15) in Nembach (1966)

LABUSCH, R., (1967) Phys. Stat. Sol. $\underline{19}$, 715

LABUSCH, R. (1968) Phys. Rev. $\underline{170}$, 470

LABUSCH, R. (1969a) Crystal Lattice Defects $\underline{1}$, 1

LABUSCH, R. (1969b) Phys. Stat. Sol. $\underline{32}$, 439

LABUSCH, R. (1974) personal communication

LABUSCH, R., and HAASEN, P. (1973) personal communication

LAVERICK, C. (1965) Proc. Intern. Symp. Mangnet Tech. UC-28, Stanford Univ., p. 560

LE BLANC, M.A.R. (1963) Phys. Rev. Letters $\underline{11}$, 149

LIPPMANN, G. (1974) Intern. Discussion Meeting on Flux Pinning in Superconductors St. Andreasberg/Harz, Germany, Sept. 23-27

LIPPMANN, G., SCHELTEN, J., HENDRICKS, R.W., and SCHMATZ, W. (1973) Phys. Stat. Sol. (b) $\underline{58}$, 633

LIPPMANN, G., SCHELTEN, J., SCHMATZ, W., and ULLMAIER, H. (1975) to be published

LIVINGSTON, J.D. (1966) Appl. Phys. Letters $\underline{8}$, 319

LONDON, H. (1962) Phys. Letters $\underline{6}$, 162

LONDON, F. (1950) Superfluids, Vol. 1, Wiley, New York

LOWELL, J. (1967) Phys. Letters $\underline{A26}$, 111

LOWELL, J. (1972) J. Phys. F $\underline{2}$, 547

MAC VICAR, M.L.A., and ROSE, R.M. (1968) J. Appl. Phys. $\underline{39}$, 721

MAILFERT, G. and PECH, T. (1972) 4th Intern. Cryogenics Eng. Conf. p. 149, IPC Science and Technology Press

MAKI, K. (1964) Physics $\underline{1}$, 21 and $\underline{1}$, 27

MAKI, K., and TSUZUKI, T. (1965), Phys. Rev. $\underline{139}$, A868

MATRICON, J. (1964) Phys. Letters $\underline{9}$, 289

MELVILLE, P.H., and TAYLOR, M.T. (1970) Cryogenics $\underline{10}$, 491

MENDELSSOHN, K. (1935) Proc. Roy. Soc. (London) $\underline{A152}$, 34

MIYAHARA, K., IRIE, F., and YAMAFUJI, K. (1969) J. Phys. Soc. Jap. $\underline{27}$, 290

MONTGOMERY, D.B. (1969) Solenoid Magnet Design, John Wiley & Sons, Inc. New York

NARLIKAR, A.V., and DEW-HUGHES, D. (1964) Phys. Stat. Sol. $\underline{6}$, 383

NEMBACH, E. (1966) Phys. Stat. Sol. $\underline{13}$, 543

NEMBACH, E., and TACHIKAWA, K. (1969) J. Less Common Metals $\underline{19}$, 359

NEWHOUSE, V.L. (1969) in: Superconductivity, Vol. II (Ed. R.D. Parks) M. Dekker, Inc., New York, p. 1283

OBST, B. (1971) Phys. Stat. Sol. (b) $\underline{45}$, 467

OROWAN, E. (1940) Proc. Phys. Soc. $\underline{52}$, 8

OTTER, F.A., and SOLOMON, P.R. (1966) Phys. Rev. Letters $\underline{16}$, 681

PARKIN, D.M., and SCHWEITZER, D.G. (1973) Nucl. Technology $\underline{22}$, 108

PEACH, M.O., and KOEHLER, J.S. (1950) Phys. Rev. $\underline{80}$, 436

PETERMANN, J. (1970) Z. Metallk. $\underline{61}$, 724

PIPPARD, A.B. (1969) Phil. Mag. $\underline{19}$, 217

PULVER, M. (1972) Phys. Stat. Sol. (b) $\underline{49}$, K95

ROSE-INNES, A.C., and STANGHAM, E.A. (1969) Cryogenics <u>9</u>, 456

ROSNER, C.H. (1974) personal communication

ST. JAMES, D., SARMA, G. and THOMAS, E.J. (1969) Type II Super-conductivity, Pergamon, Oxford

SCANLAN, R.M. (1974) to be published in J. Appl. Phys.

SCHELTEN, J., ULLMAIER, H., and SCHMATZ, W. (1971) Phys. Stat. Sol. (b) <u>48</u>, 619

SCHELTEN, J., ULLMAIER, H., and LIPPMANN, G. (1972) Z. Physik <u>253</u>, 219

SCHELTEN, J., LIPPMANN, G., and ULLMAIER, H. (1974) J. Low Temp. Phys. <u>14</u>, 213

SCHELTEN, J., ULLMAIER, H., and LIPPMANN, G. (1975) Phys. Rev. B <u>12</u>, No. 5

SCHMID, A. (1966) Phys. Kond. Mat. <u>5</u>, 302

SCHMUCKER, R. (1974) Intern. Discussion Meeting on Flux Pinning in Superconductors, St. Andreasberg/Harz, Germany, Sept. 23-27

SEEGER, A. (1968) Comments Solid State Phys. <u>1</u>, 134

SEEGER, A., and KRONMÜLLER, H. (1968) Phys. Stat. Sol. <u>27</u>, 371

SEEGER, A., SCHUMACHER, D., SCHILLING, W., and DIEHL, J. (editors) Vacancies and Interstitials in Metals, North Holland, Amsterdam, 1970

SEKULA, S.T., BOOM, R.W., and BERGERON, D.J. (1963) Appl. Phys. Letters <u>2</u>, 102

SHUBNIKOV, L.V., KHOTKEVICH, V.I., SHEPELEV, Yu.D., and RIABININ, Yu.N. (1937) Zh. Eksperim. i Teor. Fiz. <u>7</u>, 221

SIKORA, A. (1968) Phys. Letters <u>A27</u>, 175

SMITH, P.F. (1967) Proc. 2nd Intern. Conf. Magnet Technology, Oxford, Ed. H. Hadley (Rutherford Lab) p. 543

SMITH, P.F., WILSON, M.N., and SPURWAY, A.H. (1970) J. Phys. D $\underline{3}$, 1561

STEPHEN, M.J. (1966) Phys. Rev. Letter $\underline{16}$, 801

SUENAGA, M., and SAMPSON, W.B. (1971) Appl. Phys. Letters $\underline{18}$, 584

SWEEDLER, A.R., SCHWEITZER, D.G., and WEBB, G.W. (1974) Phys. Rev. Letters $\underline{33}$, 168

TACHIKAWA, K. (1972) Proc. Applied Superconductivity Conf., May 1-3, 1972, p. 371, IEEE, New York

THOMPSON, J., and JOINER, W.C.H. (1974) to be published

THOREL, P., KAHN, R., SIMON, Y., and CRIBIER, D. (1973) J. Physique $\underline{34}$, 447

TINKHAM, M. (1964) Phys. Rev. Letters $\underline{13}$, 804

TOTH, L.E., and PRATT, I.P. (1964) Appl. Phys. Letters $\underline{4}$, 75

TRÄUBLE, H., and ESSMANN, U. (1968a) Phys. Stat. Sol. $\underline{25}$, 373

TRÄUBLE, H., and ESSMANN, U. (1968b) J. Appl. Phys. $\underline{39}$, 4052

ULLMAIER, H. (1966) Phys. Stat. Sol. $\underline{17}$, 631

ULLMAIER, H. (1970) in: Festkörperprobleme, Vol. 10 (Ed. O. Madelung) Vieweg, Braunschweig, p. 367

ULLMAIER, H. (1973) Proc. Internat. Conf. Defects and Defect Clusters in BCC Metals and Their Alloys, Gaithersburg, Maryland, Aug. 1973, p. 363

ULLMAIER, H., and GAUSTER, W.F. (1966) J. Appl. Phys. $\underline{37}$, 4519

ULLMAIER, H., and KERNOHAN, R.H. (1966) Phys. Stat. Sol. $\underline{17}$, K233

ULLMAIER, H., and KERNOHAN, R.H. (1970) unpublished

ULLMAIER, H., PAPASTAIKOUDIS, C., TAKACS, S., and SCHILLING, W., (1970) Phys. Stat. Sol. $\underline{41}$, 671

ULLMAIER, H., ZELLER, R., and DEDERICHS, P.H. (1973) Phys. Letters $\underline{44A}$, 331

VAN DER KLEIN, C.A.M., KES, P.H., VAN BEELEN, H., and DE KLERK, D. (1974) J. Low Temp. Phys. 16, 169

VAN OOIJEN, D.J. (1965) Phys. Letters 14, 95 and 17, 230

VAN GURP, G.J. (1968) Phys. Rev. 166, 436 and (1969) 178, 650

VAN GURP, G.J., and VAN OOIJEN, D.J. (1966a) J. Phys. C, 27, 3

VAN GURP, G.J., and VAN OOIJEN, D.J. (1966b) J. Phys. Radium 27, C3-51

VAN VIJFEIJKEN, A.G., and NIESSEN, A.K. (1965) Phys. Letters 16, 23

VOIGT, H. (1968) Z. Physik 213, 119

WEBB, W.W. (1963) Phys. Rev. Letters 11, 191

WEBER, H.W. (1974) J. Low Temp. Phys. 17, 49

WEBER, H.W., PFEIFFER, K., and RAUCH, H. (1971) Z. Physik 244, 383

WEBER, H., and RIEGLER, R. (1973) Solid State Comm. 12, 121

WILSON, M.N., and WALTERS, C.R. (1970) J. Phys. D 3, 1547

WILSON, M.N., WALTERS, C.R., LEWIN, J.D., and SMITH, P.F. (1970) J. Phys. D 3, 1517

WIPF, S.L. (1964) Westinghouse Sci. Paper 64-110-280

WIPF, S.L. (1968) Proc. Brookhaven Summer Study on Superconductivity Devices, p. 511

WOHLLEBEN, K. (1973) J. Low Temp. Phys. 13, 269

WRAIGHT, P.C. (1971) Phil. Mag. 23, 1261

YAMAFUJI, K., and IRIE, F. (1967) Physics Letter A25, 387

YAMAMOTO, M., OHTA, N., and OHTSUKA, T. (1974) J. Low Temp. Phys. 15, 231

ZERWECK, G. (1973) Phil. Mag. 27, 197

**Springer Tracts
in Modern Physics**

Recent volumes

**Springer-Verlag
Berlin
Heidelberg
New York**

All titles with Classified Index

Zeitschrift für Physik A
Atoms and Nuclei

Editorial Board: P. Armbruster, G. Backenstoss,
I. Bergström, J. Christiansen, O. Haxel, M. Lefort,
G. zu Putlitz, L. Spruch
Managing Editor: H.A. Weidenmüller
Editor-in-Chief of Sections A and B: O. Haxel

Zeitschrift für Physik B
Condensed Matter and Quanta
Combined with "Physics of Condensed Matter"

Editorial Board: W. Brenig, W. Buckel,
R.A. Cowley, D. Cribier, L. Genzel, W. Klose,
G. Leibfried, T. Riste, T. Springer, H. Thomas,
Y. Yacoby, J. Zittartz
Managing Editor: H. Horner

Applied Physics

Editorial Board: A. Benninghoven, R. Gomer,
F. Kneubühl, H.K.V. Lotsch, H.J. Queisser,
F.P. Schäfer, A. Seeger, K. Shimoda, T. Tamir,
H.P.J. Wijn, H. Wolter

Sample copies as well as subscription and back-
volume information available upon request

Please address:

Springer-Verlag
Werbeabteilung 4021
D 1000 Berlin 33
Heidelberger Platz 3

or

Springer-Verlag
New York Inc.
Promotion Department
175 Fifth Avenue
New York, N.Y. 10010

Springer-Verlag
Berlin
Heidelberg
New York